W9-BPK-350

JUNK DNA

JUNK DNA

A Journey Through the
Dark Matter of the Genome

NESSA CAREY

COLUMBIA UNIVERSITY PRESS
NEW YORK

Columbia University Press
Publishers Since 1893
New York Chichester, West Sussex
cup.columbia.edu

Published simultaneously in the United Kingdom by Icon Books Ltd.

ISBN 978-0-231-17084-0 (cloth : alk. paper)
ISBN 978-0-231-53941-8 (e-book)
Library of Congress Control Number : 2014955417

∞

Columbia University Press books are printed on permanent
and durable acid-free paper.
This book is printed on paper with recycled content.
Printed in the United States of America

c 10 9 8 7 6 5 4 3 2 1

Cover design by Edward Bettison
Illustration by Edward Bettison

For Abi Reynolds, who is always by my side
And for Sheldon – good to see you again

Contents

Acknowledgments

I am lucky that for my second book I continue to have the support of a great agent, Andrew Lownie, and of lovely publishers. At Icon Books I'd particularly like to thank Duncan Heath, Andrew Furlow and Robert Sharman, but not forgetting their former colleagues Simon Flynn and Henry Lord. At Columbia University Press I'm very grateful to Patrick Fitzgerald, Bridget Flannery-McCoy and Derek Warker.

As always, entertainment and enlightenment have been obtained from some unusual quarters. Conor Carey, Finn Carey and Gabriel Carey all played a role in this, and outside the genetic clan I'd also like to thank Iona Thomas-Wright. Endless support and lots of biscuits have been provided by my ever-patient, delightful mother-in-law, Lisa Doran.

I've had a blast delivering lots of science talks to non-specialist audiences since my first book was published. The various organisations that have invited me to speak are too many to namecheck but they know who they are and I've enjoyed the privilege immensely. It's been very inspiring. Thank you all.

And finally Abi. Who is mercifully forgiving of the fact that, despite my promises, I still haven't had that ballroom dancing lesson yet.

Notes on Nomenclature

There's a bit of a linguistic difficulty in writing a book on junk DNA, because it is a constantly shifting term. This is partly because new data change our perception all the time. Consequently, as soon as a piece of junk DNA is shown to have a function, some scientists will say (logically enough) that it's not junk. But that approach runs the risk of losing perspective on how radically our understanding of the genome has changed in recent years.

Rather than spend time trying to knit a sweater with this ball of fog, I have adopted the most hard-line approach. Anything that doesn't code for protein will be described as junk, as it originally was in the old days (second half of the twentieth century). Purists will scream, and that's OK. Ask three different scientists what they mean by the term 'junk', and we would probably get four different answers. So there's merit in starting with something straightforward.

I also start by using the term 'gene' to refer to a stretch of DNA that codes for a protein. This definition will evolve through the course of the book.

After my first book *The Epigenetics Revolution* was published, I realised the readership was quite binary with respect to gene names. Some people love knowing which gene is being discussed, but for other readers it disrupts the flow horribly. So this time I have only used specific gene names in the text where absolutely necessary. But if you want to know them, they are in the footnotes, and the citations for the original references are at the back of the book.

JUNK DNA

An Introduction to Genomic Dark Matter

Imagine a written script for a play, or film, or television programme. It is perfectly possible for someone to read a script just as they would a book. But the script becomes so much more powerful when it is used to produce something. It becomes more than just a string of words on a page when it is spoken aloud, or better yet, acted.

DNA is rather similar. It is the most extraordinary script. Using a tiny alphabet of just four letters it carries the code for organisms from bacteria to elephants, and from brewer's yeast to blue whales. But DNA in a test tube is pretty boring. It does nothing. DNA becomes far more exciting when a cell or an organism uses it to stage a production. The DNA is used as the code for creating proteins and these proteins are vital for breathing, feeding, getting rid of waste, reproducing and all the other activities that characterise living organisms.

Proteins are so important that in the twentieth century scientists used them to define what they meant by a gene. A gene was described as a sequence of DNA that codes for a protein.

Let's think about the most famous scriptwriter in history, William Shakespeare. It can take a while for us to tune in to Shakespeare's writings because of the way the English language has changed in the centuries since his death. But even so, we are always confident that the bard only wrote the words he needed his actors to speak.

Shakespeare did not, for example, write the following:

vjeqriugfrhbvruewhqoerahcxnqowhvgbutyunyhewq
icxhjafvurytnpemxoqp[etjhnuvrwwwebcxewmoipzo
wqmroseuiednrcvtycuxmqpzjmoimxdcnibyrwvyteb
anyhcuxqimokzqoxkmdcifwrvjhentbubygdecftywer
ftxunihzxqwemiuqwjiqpodqeotherpowhdymrxname
hnfeicvbrgytrchguthhhhhhgcwouldupaizmjdpq
smellmjzufernnvgbyunasechuxhrtgcnionytuiongdjsi
oniodefnionihyhoniosdreniokikiniourvjcxoiqweopap
qsweetwxmocviknoitrbiobeierrrrrruorytnihgfiwosw
akxdcjdrfuhrqplwjkdhvmogmrfbvhncdjiwemxsklowe

Instead, he just wrote the words which are underlined:

vjeqriugfrhbvruewhqoerahcxnqowhvgbutyunyhewq
icxhjafvurytnpemxoqp[etjhnuvrwwwebcxewmoipzo
wqmroseuiednrcvtycuxmqpzjmoimxdcnibyrwvyteb
anyhcuxqimokzqoxkmdcifwrvjhentbubygdecftywer
ftxunihzxqwemiuqwjiqpodqeotherpowhdymrxname
hnfeicvbrgytrchguthhhhhhgcwouldupaizmjdpq
smellmjzufernnvgbyunasechuxhrtgcnionytuiongdjsi
oniodefnionihyhoniosdreniokikiniourvjcxoiqweopap
qsweetwxmocviknoitrbiobeierrrrrruorytnihgfiwosw
akxdcjdrfuhrqplwjkdhvmogmrfbvhncdjiwemxsklowe

That is, 'A rose by any other name would smell as sweet'.

But if we look at our DNA script it is not sensible and compact, like Shakespeare's line. Instead, each protein-coding region is like a single word adrift in a sea of gibberish.

For years, scientists had no explanation for why so much of our DNA doesn't code for proteins. These non-coding parts were dismissed with the term 'junk DNA'. But gradually this position has begun to look less tenable, for a whole host of reasons.

Perhaps the most fundamental reason for the shift in emphasis is the sheer volume of junk DNA that our cells contain. One of the biggest shocks when the human genome sequence was completed in 2001 was the discovery that over 98 per cent of the DNA in a human cell is junk. It doesn't code for any proteins. The Shakespeare analogy used above is in fact a simplification. In genome terms, the ratio of gibberish to text is about four times as high as shown. There are over 50 letters of junk for every one letter of sense.

There are other ways of envisaging this. Let's imagine we visit a car factory, perhaps for something high-end like a Ferrari. We would be pretty surprised if for every two people who were building a shiny red sports car, there were another 98 who were sitting around doing nothing. This would be ridiculous, so why would it be reasonable in our genomes? While it's a very fair point that it's the imperfections in organisms that are often the strongest evidence for descent from common ancestors – we humans really don't need an appendix – this seems like taking imperfection rather too far.

A much more likely scenario in our car factory would be that for every two people assembling a car, there are 98 others doing all the things that keep a business moving. Raising finance, keeping accounts, publicising the product, processing the pensions, cleaning the toilets, selling the cars etc. This is probably a much better model for the role of junk in our genome. We can think of proteins as the final end points required for life, but they will never be properly produced and coordinated without the junk. Two people can build a car, but they can't maintain a company selling it, and certainly can't turn it into a powerful and financially successful brand. Similarly, there's no point having 98 people mopping the floors and staffing the showrooms if there's nothing to sell. The whole organisation only works when all the components are in place. And so it is with our genomes.

The other shock from the sequencing of the human genome was the realisation that the extraordinary complexities of human anatomy, physiology, intelligence and behaviour cannot be explained by referring to the classical model of genes. In terms of numbers of genes that code for proteins, humans contain pretty much the same quantity (around 20,000) as simple microscopic worms. Even more remarkably, most of the genes in the worms have directly equivalent genes in humans.

As researchers deepened their analyses of what differentiates humans from other organisms at the DNA level, it became apparent that genes could not provide the explanation. In fact, only one genetic factor generally scaled with complexity. The only genomic features that increased in number as animals became more complicated were the regions of junk DNA. The more sophisticated an organism, the higher the percentage of junk DNA it contains. Only now are scientists really exploring the controversial idea that junk DNA may hold the key to evolutionary complexity.

In some ways, the question raised by these data is pretty obvious. If junk DNA is so important, what is it actually doing? What is its role in a cell, if it isn't coding for proteins? It's becoming apparent that junk DNA actually has a multiplicity of different functions, perhaps unsurprisingly given how much of it there is.

Some of it forms specific structures in the chromosomes, the enormous molecules into which our DNA is packaged. This junk prevents our DNA from unravelling and becoming damaged. As we age, these regions decrease in size, finally declining below a critical minimum. After that, our genetic material becomes susceptible to potentially catastrophic rearrangements that can lead to cell death or cancers. Other structural regions of junk DNA act as anchor points when chromosomes are shared equally between different daughter cells during cell division. (The term 'daughter cell' means any cell created by division of a parental cell. It doesn't imply that the cell is female.) Yet others act as insulation regions, restricting gene expression to specific regions of chromosomes.

But a great deal of our junk DNA is not simply structural. It doesn't code for proteins, but it does code for a different type of molecule, called RNA. A large class of this junk DNA forms factories in the cell, helping to produce proteins. Other types of RNA molecules transport the raw material for protein production to the factory sites.

Other regions of junk DNA are genetic interlopers, derived from the genomes of viruses and other microorganisms that have integrated into human chromosomes, like genetic sleeper agents. These remnants of long-dead organisms carry potential dangers to the cell, the individual and sometimes even to wider populations. Mammalian cells have developed multiple mechanisms to keep these viral elements silent, but these systems can break down. When they do, the effects can range from relatively benign – changing the coat colour of a particular strain of mice – to much more dramatic, such as an increased risk of cancer.

A major role of junk DNA, only recognised in the main in the last few years, is to regulate gene expression. Sometimes this can have a huge and noticeable effect in an individual. One particular piece of junk DNA is absolutely vital for ensuring healthy gene expression patterns in female animals. Its effects are seen in a whole range of situations. A mundane example is the control of the colour patterns of tortoiseshell cats. At its most extreme, the same mechanism also explains why female identical twins may present with different symptoms of a genetically inherited disease. In some cases, this can be so extreme that one twin is severely affected with a life-threatening disorder while the other is completely healthy.

Thousands and thousands of regions of junk DNA are suspected to regulate networks of gene expression. They act like the stage directions for the genetic script, but directions of a complexity we could never envisage in the theatre. Forget about 'Exit, pursued by a bear'. These would be more along the lines of 'If performing *Hamlet* in Vancouver and *The Tempest* in Perth, then put the stress on the fourth syllable of this line of *Macbeth*. Unless there's an amateur production of *Richard III* in Mombasa and it's raining in Quito.'

Researchers are only just beginning to unravel the subtleties and interconnections in the vast networks of junk DNA. The field is controversial. At one extreme we have scientists claiming experimental proof is lacking to support sometimes sweeping claims. At the other are those who feel there is a whole generation of scientists (if not more) trapped in an outdated model and unable to see or understand the new world order.

Part of the problem is that the systems we can use to probe the functions of junk DNA are still relatively underdeveloped. This can sometimes make it hard for researchers to use experimental approaches to test their hypotheses. We have only been working on this for a relatively short space of time. But sometimes we need to remember to step back from the lab bench and the machines that go ping. Experiments surround us every day, because nature and evolution have had billions of years to try out all sorts of changes. Even the brief geological moment that represents the emergence and spread of our own species has been sufficient time to create a greater range of experiments than those of us who wear lab coats could ever dream of testing. Consequently, throughout much of this book we will explore the darkness by using the torch of human genetics.

There are many ways to begin shining a light on the dark matter of our genome, so let's start with an odd but unassailable fact to anchor us. Some genetic diseases are caused by mutations in junk DNA, and there is probably no better starting point for our journey into the hidden genomic universe than this.

1. Why Dark Matter Matters

Sometimes life seems to be cruel in the troubles it piles onto a family. Consider this example. A baby boy was born; let's call him Daniel. He was strangely floppy at birth, and had trouble breathing unassisted. With intensive medical care Daniel survived and his muscle tone improved, allowing him to breathe unaided and to develop mobility. But as he grew older it became apparent that Daniel had pronounced learning disabilities that would hold him back throughout life.

His mother Sarah loved Daniel and cared for him every day. As she entered her mid-30s this became more difficult because Sarah developed strange symptoms. Her muscles became very stiff, to the extent that she would have trouble releasing items after grasping them. She had to give up her highly skilled part-time job as a ceramics restorer. Her muscles also began to waste away noticeably. Yet she found ways to cope. But when she was only 42 years old Sarah died suddenly from a cardiac arrhythmia, a catastrophic disruption in the electrical signals that keep the heart beating in a coordinated way.

It fell to Sarah's mother, Janet, to look after Daniel. This was challenging for her, and not just because of her grandson's difficulties and the grief she was suffering over the early death of her daughter. Janet had developed cataracts in her early 50s and as a consequence her vision wasn't that great.

It seemed as if the family had suffered a very unfortunate combination of unrelated medical problems. But specialists began to notice something rather unusual. This pattern – cataracts in one individual, muscle stiffness and cardiac defects in their daughter and floppy muscles and learning disabilities in the grandchildren – occurred in multiple families. These individual families lived all over the world and none of them were related to each other.

Scientists realised they were looking at a genetic disease. They named it myotonic dystrophy (myotonic means muscle tone, dystrophy means wasting). The condition occurred in every generation of an affected family. On average there was a one in two chance of a child being affected if their parent had the condition. Males and females were equally at risk and either could pass it on to their children.[1]

These inheritance characteristics are very typical of diseases caused by mutations in a single gene. A mutation is simply a change from the normal DNA sequence. We typically inherit two copies of every gene in our cells, one from our mother and one from our father. The pattern of inheritance in myotonic dystrophy, where the disease appears in each generation, is referred to as dominant. In dominant disorders, only one of the two copies of a gene carries the mutation. It is the copy inherited from the affected parent. This mutated gene is able to cause the disease even though the cells also contain a normal copy. The mutated gene somehow 'dominates' the action of the normal gene.

But myotonic dystrophy also had characteristics that were very different from a typical dominant disorder. For a start, dominant disorders don't normally get worse as they are passed on from parent to child. There is no reason why they should, because the affected child inherits the same mutation as the affected parent. Patients with myotonic dystrophy also developed symptoms at earlier ages as the disorder was passed on down the generations, which again is unusual.

There was another way in which myotonic dystrophy was different from the normal genetic pattern. The severe congenital form of the disease, the one that affected Daniel, was only ever found in the children of affected mothers. Fathers never passed on this really severe form.

In the early 1990s a number of different research groups identified the genetic change that causes myotonic dystrophy. Fittingly for an unusual disease, it was a very unusual mutation. The myotonic dystrophy gene contains a small sequence of DNA that is repeated multiple times.[2] The small sequence is made from three of the four 'letters' that make up the genetic alphabet used by DNA. In the myotonic dystrophy gene, this repeated sequence is formed by the letters C, T and G (the other letter in the genetic alphabet is A).

In people without the myotonic dystrophy mutation, there can be anything from five to around 30 copies of this CTG motif, one after the other. Children inherit the same number of repeats as their parents. But when the number of repeats gets larger, greater than 35 or thereabouts, the sequence becomes a bit unstable and may change in number when it is passed on from parent to child. Once it gets above 50 copies of the motif, the sequence becomes really unstable. When this happens, parents can pass on much bigger repeats to their children than they themselves possess. As the repeat length increases, the symptoms become more severe and are obvious at an earlier age. That's why the disease gets worse as it passes down the generations, such as in the family that opened this chapter. It also became apparent that usually only mothers passed on the really big repeats, the ones that led to the severe congenital phenotype.

This ongoing expansion of a repeated sequence of DNA was a very unusual mutation mechanism. But the identification of the expansion that causes myotonic dystrophy shone a light on something even more unusual.

Knitting with DNA

Until quite recently, mutations in gene sequences were thought to be important not because of the change in the DNA itself but because of their downstream consequences. It's a little like a mistake in a knitting pattern. The mistake doesn't matter when it's just a notation on a piece of paper. The mistake only becomes a problem when you knit something and end up with a hole in your sweater or three sleeves on your cardigan because of the error in the knitting code.

A gene (the knitting pattern) ultimately codes for a protein (the sweater). It's proteins that we think of as the molecules in our cells that do all the work. They carry out an enormous number of functions. These include the haemoglobin in our red blood cells that carries oxygen around our bodies. Another protein is insulin, which is released from the pancreas to encourage muscle cells to take in glucose. Thousands and thousands of other proteins carry out the dizzying range of functions that underlie life.

Proteins are made from building blocks called amino acids. Mutations generally change the sequence of these amino acids. Depending on the mutation and where it lies in the gene, this can lead to a number of consequences. The abnormal protein may carry out the wrong function in a cell, or may not be able to work at all.

But the myotonic dystrophy mutation doesn't change the amino acid sequence. The mutated gene still codes for exactly the same protein. It was incredibly difficult to understand how the mutation led to a disease, when there was nothing wrong with the protein.

It would be tempting to write off the myotonic dystrophy mutation as some bizarre outlier with no impact for the majority of biological circumstances. That way we could put it to one side and forget about it. But it's not alone.

Fragile X syndrome is the commonest form of inherited learning disability. Mothers don't usually have any symptoms but they pass the condition on to their sons. The mothers carry the mutation but are not affected by it. Like myotonic dystrophy, this disorder is also caused by increases in the length of a three-letter sequence. In this case, the sequence is CCG. And just like myotonic dystrophy, this increase doesn't change the sequence of the protein encoded by the Fragile X gene.

Friedreich's ataxia is a form of progressive muscle wasting in which symptoms normally appear in late childhood or early adolescence. In contrast to myotonic dystrophy, the parents are usually unaffected by the disorder. Both the mother and father are carriers. Each parent possesses one normal and one abnormal copy of the relevant gene. But if a child inherits a mutated copy from each parent, the child develops the disease. Friedreich's ataxia is also caused by an increase in a three-letter sequence, GAA in this case. And once again it doesn't change the sequence of the protein encoded by the affected gene.[3]

These three genetic diseases, so different in their family histories, symptoms and inheritance patterns, nevertheless told scientists something quite consistent: there are mutations that can cause disease without changing the amino acid sequence of proteins.

An impossible disease

An even more startling discovery was made a few years later. There is another inherited wasting disorder in which the muscles of the face, shoulders, and upper arms gradually weaken and degenerate. The disease is named after this pattern – it's called facioscapulohumeral muscular dystrophy. Perhaps unsurprisingly, this is usually shortened to FSHD. Symptoms are usually detectable by the time a patient is in their early 20s. Like myotonic dystrophy, the disease is dominant and passed from affected parent to child.[4]

Scientists spent years looking for the mutation that causes FSHD. Eventually, they tracked it down to a repeated DNA sequence. But in this case the mutation is very different from the three-letter repeats found in myotonic dystrophy, fragile X syndrome and Friedreich's ataxia. It is a stretch of over 3,000 letters. We can call this a block. In people who don't suffer from FSHD, there are from eleven to about 100 blocks, one after another. But patients with FSHD have a small number of blocks, ten at most.

That was unexpected. But the real shock for the researchers was that they really struggled to find a gene near the mutation.

Genetic diseases have given us great new insights into biology over the last hundred years or so. It's easy to underestimate how hard-won some of that knowledge was. The identification of the mutations described here usually represented over a decade of work for significant numbers of people. It was entirely dependent on access to families who were willing to give blood samples and trace their family histories to help scientists home in on the key individuals to analyse.

The reason this kind of analysis was so difficult was because researchers were normally looking for a very small change in a very large landscape, hunting for a single specific acorn in a forest. This all became much easier from 2001 onwards, after the release of the human genome sequence. The genome is the entire sequence of DNA in our cells.

Because of the Human Genome Project, we know where all the genes are positioned relative to one another, and their sequences. This, together with enormous improvements in the technologies used to sequence DNA, has made it much faster and cheaper to find the mutations underlying even very rare genetic diseases.

But the completion of the human genome sequence has had impact far beyond identifying the mutations that cause disease. It's changing many of our ideas about some of the most fundamental ideas that have held sway in biology since we first understood that DNA was our genetic material.

When considering how our cells work, almost every scientist over the last six decades has been focused on the impacts of proteins. But from the moment the human genome was sequenced, scientists have had to face a rather puzzling dilemma. If proteins are so all-important, why is only 2 per cent of our DNA devoted to coding for amino acids, the building blocks of proteins? What on earth is the other 98 per cent doing?

2. When Dark Matter Turns Very Dark Indeed

The astonishing percentage of the genome that didn't code for proteins was a shock. But it was the scale of the phenomenon that was surprising, not the phenomenon itself. Scientists had known for many years that there were stretches of DNA that didn't code for proteins. In fact, this was one of the first big surprises after the structure of DNA itself was revealed. But hardly anyone anticipated how important these regions would prove to be, nor that they would provide the explanation for certain genetic diseases.

At this point it's worth looking in a little more detail at the building blocks of our genome. DNA is an alphabet, and a very simple one at that. It is formed of just four letters – A, C, G and T. These are also known as bases. But because our cells contain so much DNA, this simple alphabet carries an incredible amount of information. Humans inherit 3 billion of the bases that make up our genetic code from our mother, and a similar set from our father. Imagine DNA as a ladder, with each base representing a rung, and each rung being 25cm from the next. The ladder would stretch 75 million kilometres, roughly from earth to Mars (depending on the relative positions of their orbits on the day the ladder was put in place).

To think of it another way, the complete works of Shakespeare are reported to contain 3,695,990 letters.[1] This means we inherit the equivalent of just over 811 books the length of the Bard's canon from mum and the same number from dad. That's a lot of information.

If we extend our alphabet analogy a bit further, the DNA alphabet encodes words of just three letters each. Each three-letter word acts as the placeholder for a specific amino acid, the building blocks of proteins. A gene can be thought of as a sentence of three-letter words, which acts as the code for a sequence of amino acids forming a protein. This is summarised in Figure 2.1.

Each cell usually contains two copies of any given gene. One was inherited from the mother and one from the father. But although there are only two copies of each gene in a cell, that same cell can create thousands and thousands of the protein molecules encoded by a specific gene.

This is because there are two amplification mechanisms built into gene expression. The sequence of bases in the DNA doesn't act as the direct template for the protein. Instead, the cell makes copies of the gene. These copies are very similar to the DNA gene itself, but not identical. They have a slightly different chemical composition and are known as RNA (ribonucleic acid, instead of the deoxyribonucleic acid in DNA). Another difference is that in RNA, the base T is replaced by the base U. DNA is formed of two strands joined together via pairs of bases. We could visualise this as looking a little like a railway track. The two rails are held

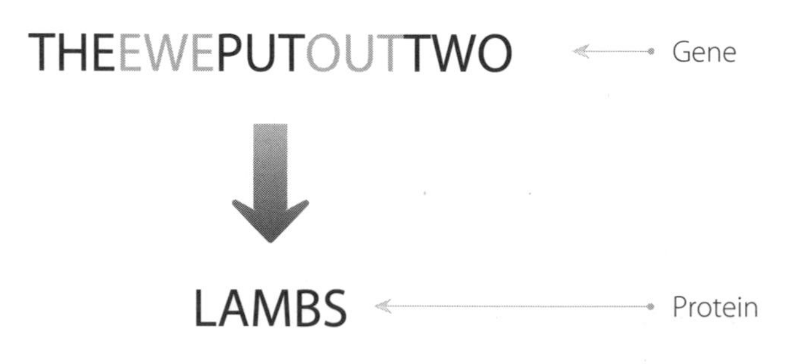

Figure 2.1 The relationship between a gene and a protein. Each three-letter sequence in the gene codes for one building block in the protein.

together by a base on one rail linking to a base on the other, as if the bases were holding hands. They only link up in a set pattern. T holds hands with A, C holds hands with G. Because of this arrangement, we tend to refer to DNA in terms of base pairs. RNA is a single-stranded molecule, just one rail. The key differences between DNA and RNA are shown in Figure 2.2. A cell can make thousands of RNA copies of a DNA gene really quickly, and this is the first amplification step in gene expression.

The RNA copies of a gene are transported away from the DNA to a different part of the cell, called the cytoplasm. In this distinct region of the cell, the RNA molecules act as the placeholders for the amino acids that form a protein. Each RNA molecule can act as a template multiple times, and this introduces the second amplification step in gene expression. This is shown diagrammatically in Figure 2.3.

We can visualise this using the analogy of the knitting pattern from Chapter 1. The DNA gene is the original knitting pattern. This pattern can be photocopied multiple times, akin to producing

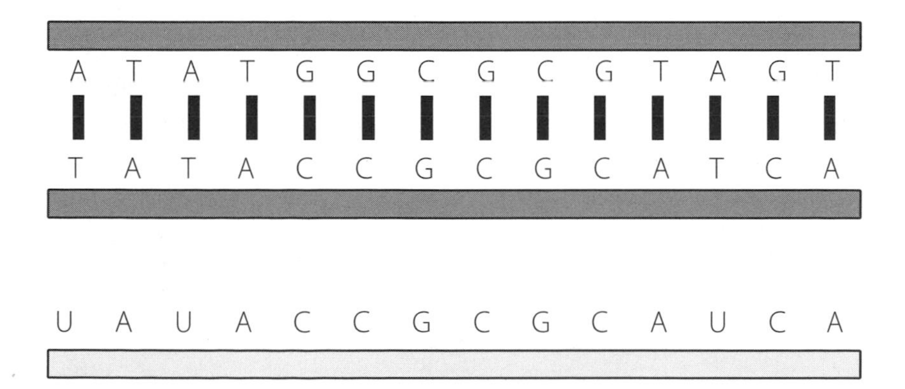

Figure 2.2 The upper panel represents DNA, which is double-stranded. The bases – A, C, G and T – hold the two strands together by pairing up. A always pairs with T, and C always pairs with G. The lower panel represents RNA, which is single-stranded. The backbone of the strand has a slightly different composition from DNA, as indicated by the different shading. In RNA, the base T is replaced by the base U.

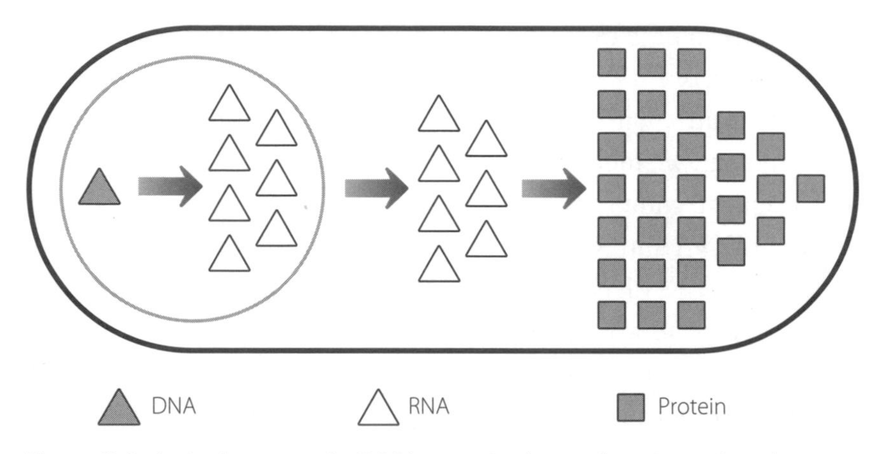

Figure 2.3 A single copy of a DNA gene in the nucleus is used as the template to create multiple copies of a messenger RNA molecule. These multiple RNA molecules are exported out of the nucleus. Each can then act as the instructions for production of a protein. Multiple copies of the same protein can be produced from each messenger RNA molecule. There are therefore two amplification steps in generating protein from a DNA code. For simplicity, only one copy of the gene is shown, although usually there will be two – one inherited from each parent.

the RNA. The copies can be sent to lots of people who can each knit the same pattern multiple times, just like creating the protein. It's a simple but efficient operating model and it works – one original pattern resulted in lots of soldiers with warm feet in the Second World War.

The RNA molecule acts as a messenger molecule, carrying a gene sequence from the DNA to the protein assembly factory. Rather logically it is therefore known as messenger RNA.

Taking out the nonsense

So far, things might seem very straightforward but scientists discovered quite some time ago that there is a strange complication. Most genes are split up into bits that code for the amino acids in a protein and intervening bits that don't. The bits that don't are like gobbledegook in the middle of a string of sensible words. These intervening bits of nonsense are known as introns.

When the cell makes RNA, it originally copies all of the DNA letters in a gene, including the bits that don't code for amino acids. But then the cell removes all the bits that don't code for protein, so that the final messenger RNA is a good instruction set for the final protein. This process is known as splicing, and Figure 2.4 shows diagrammatically how this happens.

As Figure 2.4 shows, a protein is encoded from modular blocks of information. This modularity gives the cell a lot of flexibility in how it processes the RNA. It can vary the modules which it joins together from a messenger RNA molecule, creating a range of final messengers that code for related but non-identical proteins. This is shown in Figure 2.5.

The bits of gobbledegook between the parts of a gene that code for amino acids were originally considered to be nothing but

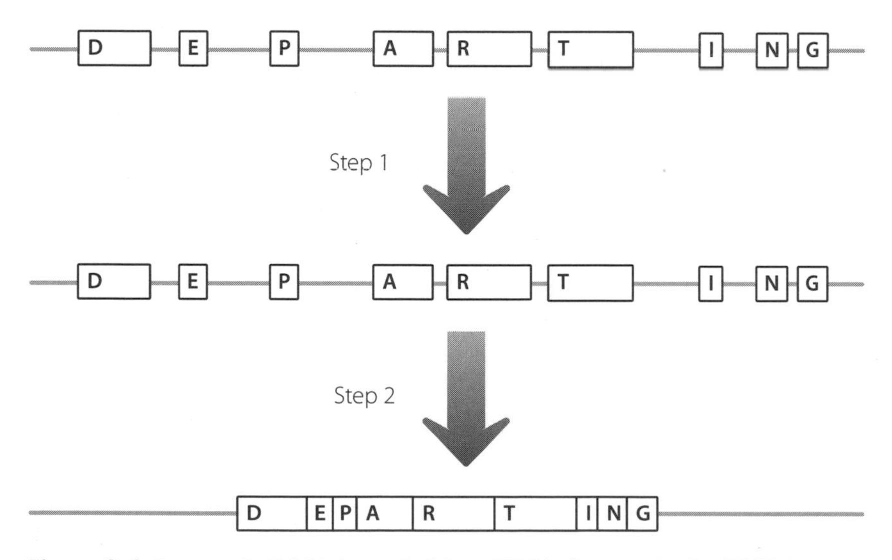

Figure 2.4 In step 1, DNA is copied into RNA. In step 2, the RNA is processed so that only the amino acid-coding regions, denoted by boxes containing letters, are joined together. The intervening junk regions are removed from the mature messenger RNA molecule.

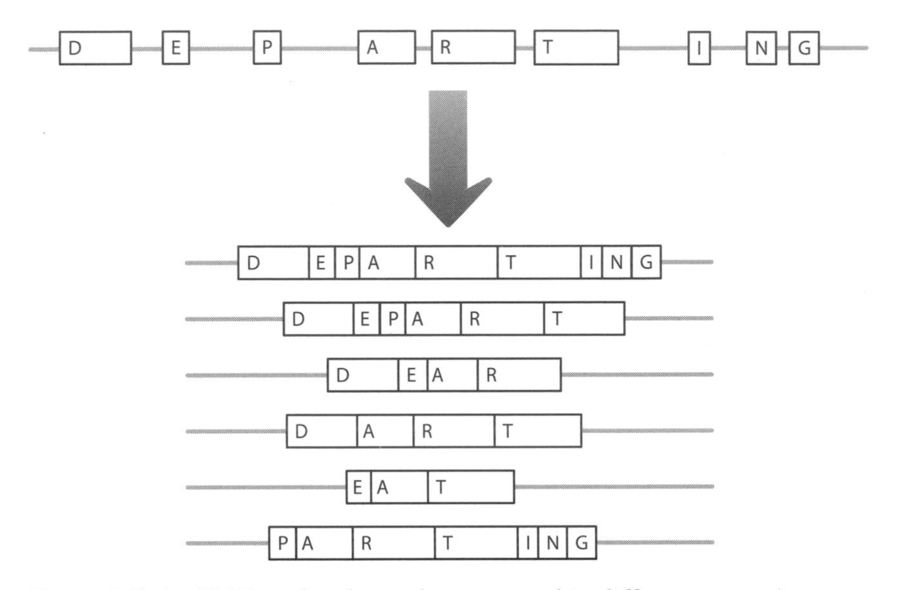

Figure 2.5 An RNA molecule can be processed in different ways. As a result, different amino acid-coding regions can be joined together. This allows different versions of a protein molecule to be produced from one original DNA gene.

nonsense or rubbish. They were referred to as junk or garbage DNA, and pretty much dismissed as irrelevant. As mentioned earlier, from here on in, we'll use the term 'junk' to denote any DNA that doesn't code for protein.

But we now know that they can have a very big impact. In Friedreich's ataxia, which we met in Chapter 1, the disorder is caused by an abnormally expanded stretch of GAA repeats in one of the junk regions, between two sections that encode amino acids. This raised the perfectly reasonable question – if the mutation doesn't affect the amino acid sequence, why do people with this mutation develop such debilitating symptoms?

The mutation in the Friedreich's ataxia gene occurs in the junk region between the first two amino acid-coding regions. In Figure 2.5, this would be between regions 'D' and 'E'. A normal gene contains from five to 30 GAA repeats but a mutated gene contains from 70 up to 1,000 repeated GAA motifs.[2] Researchers showed that when cells contained this expanded repeat, they stopped producing the messenger RNA encoded by the gene. Because they didn't make messenger RNA, they couldn't make the protein either. If you don't send out the copies of the knitting patterns, the soldiers don't get socks.

In fact, the cells didn't even make the long, unprocessed RNA copy of the gene.[3] The big GAA expansion acts as a 'sticky' region, which prevents good copying of the DNA. It's analogous to trying to photocopy a 50-page document, when pages four to twelve have been glued together. They won't feed into the copier, and the process grinds to a halt, for that particular document. In the case of the Friedreich's ataxia gene, no copying means no RNA, which means no protein.

It's not completely clear why lack of the protein encoded by the Friedreich's ataxia gene causes the disease symptoms. The protein seems to be involved in preventing iron overload in the parts of the cell that generate energy.[4] When a cell fails to produce the protein, the iron rises to toxic levels. Some cell types seem to be more sensitive than others to iron levels, and these include the ones affected in the disease.

A related but different mechanism accounts for Fragile X syndrome, the form of learning disability we encountered in Chapter 1. The mutation in Fragile X syndrome is the expansion of a CCG three-base repeat. Similarly to the Friedreich's ataxia mutation, there are usually fifteen to 65 copies of the repeat on a normal chromosome. On a chromosome carrying the Fragile X mutation there are from around 200 to several thousand copies.[5,6] But the expansion lies in a different part of the gene in Fragile X compared with Friedreich's ataxia. The mutation is found before the first amino acid-coding region, essentially in the junk to the left of block 'D' in Figure 2.5. When the junk repeat gets very large, no messenger RNA is produced, and consequently there is no protein produced from this gene.[7]

The function of the Fragile X protein is to carry lots of different RNA molecules around in the cell. This gets them to the correct locations, influences how these RNAs are processed and how they generate proteins. If there is no Fragile X protein, the other RNA molecules aren't properly regulated, and this plays havoc with the normal functioning of the cell.[8] For reasons that aren't clear, the neurons in the brain seem particularly sensitive to this effect, hence the learning disability in this disorder.

An everyday analogy may help with visualising this. In the UK, a relatively small amount of snow can incapacitate the transport networks. The snow covers the roads and the railway tracks, preventing cars and trains from moving. When this happens, people can't get to their place of work and this creates all sorts of problems. Schools can't open, deliveries aren't made, banks can't dispense cash, etc. One starting event – the snow – has all sorts of consequences because it ruins the transport systems in society. A similar thing happens in Fragile X syndrome. Just like snow on the roads and railway tracks, the effect of the mutation is to mess up a transport system in the cell, with multiple knock-on effects.

Switching off the expression of a specific gene is the key step in the pathology of both Friedreich's ataxia and Fragile X syndrome. Support for this hypothesis has been provided by very rare cases of both disorders. There are small numbers of patients where the repeat in the junk regions is of the same small size found in most healthy people. In these patients, there are mutations that change the sequence in the amino acid-coding regions. These particular amino acid sequence changes actually make it impossible for the cell to produce the protein. In other words, it doesn't matter why the protein isn't expressed. If it's not expressed, the patients have the symptoms.

Just when you have a nice theory

So far it might seem like there's a nice straightforward theme emerging. We could speculate that expansions in the junk regions are only important because they create abnormal DNA. This DNA isn't handled properly by the cells, resulting in a lack of specific important proteins. We could suggest that normally these junk regions are unimportant, with no significant role in the cell.

But there is something that argues against this. The normal range of repeats in both the Fragile X and Friedreich's ataxia genes is found in all human populations, and has been retained throughout human evolution. If these regions were completely nonsensical we would expect them to have changed randomly over time, but they haven't. This suggests that the normal repeats have some function.

But the real grit in this genetic oyster comes from myotonic dystrophy, the disorder that opened Chapter 1. The myotonic dystrophy expansion gets bigger as it passes down the generations. A parent's chromosome may contain the sequence CTG repeated 100 times, one after another. But when they pass this on to their child, this may have expanded so the child's chromosome has the sequence CTG repeated 500 times. As the number of CTG repeats gets larger, the disease becomes more and more severe. This isn't what we would expect if the expansion just switches off the nearby gene. All cells of someone with myotonic dystrophy contain two copies of the gene. One carries the normal number of repeats, and the other carries the expanded number. So, one copy of the gene should always be producing the normal amount of protein. That would mean that the most the overall levels of the protein should drop would be about 50 per cent.

We could hypothesise that as the repeat gets longer there is progressively less gene expression from the mutant version of the gene. This could lead to a gradual decline in the amount of protein produced overall. This could range from a 1 per cent drop overall for fairly small expansions, to a 50 per cent final decrease for the large ones. This could lead to different symptoms. The problem is that there aren't really any inherited genetic diseases like this. We just don't see disorders where very minor variations in expression have such a big effect (all patients with the expansion develop symptoms), but with such fine tuning between patients (the symptoms becoming more extreme as the expansion lengthens).

It's worth looking at where the expansion occurs in the myotonic dystrophy gene. It's right at the far end, after the last amino acid-coding region. In Figure 2.5, this would be on the horizontal line to the right of box 'G'. This means that the entire amino acid-coding region can be copied into RNA before the copying machinery encounters the expansion.

It's now clear that the expansion itself gets copied into RNA. It is even retained when the long RNA is processed to form the messenger RNA. The myotonic dystrophy messenger RNA does something unusual. It binds lots of protein molecules that are present in the cell. The bigger the expansion, the more protein molecules that get bound. The mutant myotonic dystrophy messenger RNA acts like a kind of sponge, mopping up more and more of these proteins. The proteins that bind to the expansion in the myotonic dystrophy messenger RNA are normally involved in regulating lots of other messenger RNA molecules. They influence how well messenger RNA molecules are transported in the cell, how long the messenger RNA molecules survive in the cell and how efficiently they encode proteins. But if all these regulators are mopped up by the expansion in the myotonic dystrophy gene messenger RNA, they aren't available to do their normal job.[9] This is shown in Figure 2.6.

Again an analogy may help. Imagine a city where every member of the police force is engaged in controlling a riot in a single location. There will be no officers left for normal policing, and burglars and car thieves may run amok elsewhere in the city. It's the

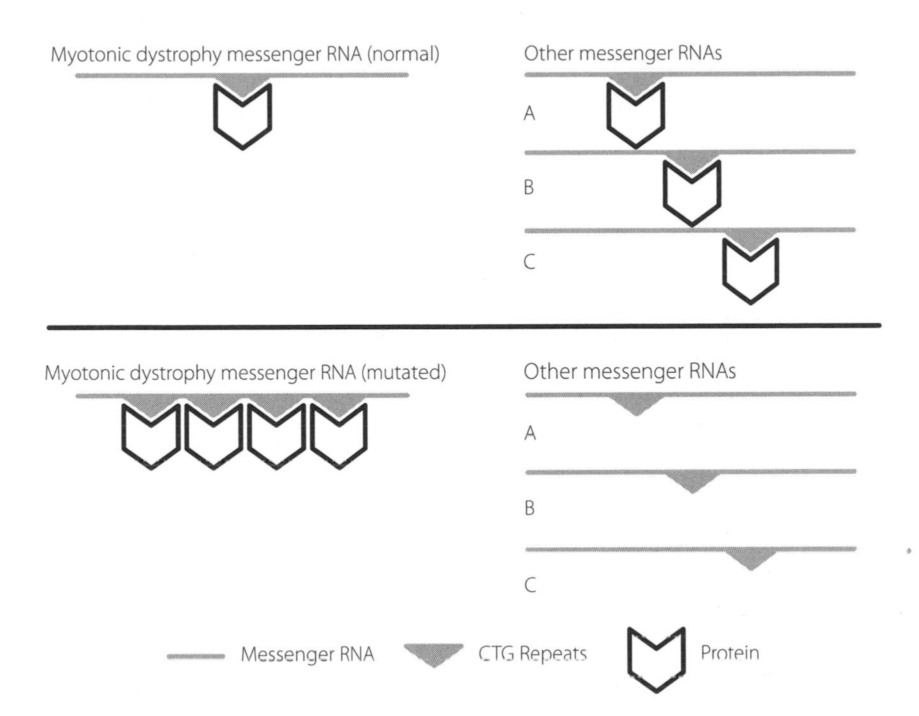

Figure 2.6 The upper panel shows the normal situation. Specific proteins, represented by the chevron, bind to the CTG repeat region on the myotonic dystrophy messenger RNA. There are plenty of these protein molecules available to bind to other messenger RNAs to regulate them. In the lower panel, the CTG sequence is repeated many times on the mutated myotonic dystrophy messenger RNA. This mops up the specific proteins, and there aren't enough left to regulate other messenger RNAs. For clarity, only a small number of repeats have been represented. In severely affected patients, they may number in the thousands.

same principle in the cells of people with the myotonic dystrophy mutation. The CTG repeat sequence expansion in a single gene – the myotonic dystrophy gene – ultimately leads to mis-regulation of a whole number of other genes in the cell.

This is because the expansion mops up more and more of the binding proteins as it gets larger. This leads to disruption of a greater quantity of other messenger RNAs, causing problems for increasing numbers of cellular functions. This eventually results in the wide range of symptoms found in patients carrying the myotonic dystrophy mutation, and explains why the patients with the largest repeats have the most severe clinical problems.

Just as we saw in Friedreich's ataxia and Fragile X syndrome, the normal CTG repeat sequences in the myotonic dystrophy gene have been highly conserved in human evolution. This is consistent with them having a healthy and important functional role. We are even more convinced this is the case for the myotonic dystrophy gene because of the proteins that bind to the repeat in the messenger RNA. These also bind to shorter repeat lengths, of the size that are present in normal genes. They just don't bind in the same abundance as they do when the repeat has expanded.

It's clear from the myotonic dystrophy example that there is a reason why messenger RNA molecules contain regions that don't code for proteins. These regions are critical for regulating how the messenger RNAs are used by the cells, and create yet another level of control, fine-tuning the amount of protein ultimately produced from a DNA gene template. But what no one appreciated when the myotonic dystrophy mutation was identified, almost ten years before the release of the human genome sequence, was just how extraordinarily complex and variable this fine-tuning would turn out to be.

3. Where Did All the Genes Go?

On 26 June 2000, it was announced that the initial draft of the sequence of the human genome had been completed. In February 2001, the first papers describing this draft sequence in detail were released. It was the culmination of years of work and technological breakthroughs, and more than a little rivalry. The National Institutes of Health in the USA and the Wellcome Trust in the UK had poured in the majority of the approximately $2.7 billion[1] required to fund the research. This was carried out by an international consortium, and the first batch of papers detailing the findings included over 2,500 authors from more than 20 laboratories worldwide. The bulk of the sequencing was carried out by five laboratories, four of them in the US and one in the UK. Simultaneously, a private company called Celera Genomics was attempting to sequence and commercialise the human genome. But by releasing their data on a daily basis as soon as it was generated, the publicly funded consortium was able to ensure that the sequence of the human genome entered the public domain.[2]

An enormous hoopla accompanied the declaration that the draft human genome had been completed. Perhaps the most flamboyant statement was from US President Bill Clinton, who declared that 'Today we are learning the language in which God created life'.[3] We can only speculate on the inner feelings of some of the scientists who had played such a major role in the project as a politician invoked a deity at the moment of technological triumph. Luckily, researchers tend to be a shy lot, especially when confronted by celebrities and TV cameras, so few expressed any disquiet publicly.

Michael Dexter was the Director of the Wellcome Trust, which had poured enormous sums of money into the Human Genome Project. He was not much less fulsome, albeit somewhat less theistic, when he defined the completion of the draft sequence as 'The outstanding achievement not only of our lifetime, but in terms of human history'.[4]

You might not be alone in thinking that perhaps other discoveries have given the Human Genome Project a run for its money in terms of impact. Fire, the wheel, the number zero and the written alphabet spring to mind, and you probably have others on your own list. It could also be claimed that the human genome sequence has not yet delivered on some of the claims that were made about how quickly it would impact on human disease. For instance, David Sainsbury, the then UK Science Minister, stated that 'We now have the possibility of achieving all we ever hoped for from medicine'.[5]

Most scientists knew, however, that these claims should be taken with whole shovelfuls of salt, because we have been taught this by the history of genetics. Consider a couple of relatively well-known genetic diseases. Duchenne muscular dystrophy is a desperately sad disorder in which affected boys gradually lose muscle mass, degenerate physically, lose mobility and typically die in adolescence. Cystic fibrosis is a genetic condition in which the lungs can't clear mucus, and the sufferers are prone to severe life-threatening infections. Although some cystic fibrosis patients now make it to the age of about 40, this is only with intensive physical therapy to clear their lungs every day, plus industrial levels of antibiotics.

The gene that is mutated in Duchenne muscular dystrophy was identified in 1987 and the one that is mutated in cystic fibrosis was identified in 1989. Despite the fact that mutations in these genes were shown to cause disease over a decade before the completion of the human genome sequence, there are still no effective treatments for these diseases after 20-plus years of trying. Clearly, there's going to be a long gap between knowing the sequence of the human genome, and developing life-saving treatments for common diseases. This is especially the case when diseases are caused by more than one gene, or by the interplay of one or more genes with the environment, which is the case for most illnesses.

But we shouldn't be too harsh on the politicians we have quoted. Scientists themselves drove quite a lot of the hype. If you are requesting the better part of $3 billion of funding from your paymasters, you need to make a rather ambitious pitch. Knowing the human genome sequence is not really an end in itself, but that doesn't make it unimportant as a scientific endeavour. It was essentially an infrastructure project, providing a dataset without which vast quantities of other questions could never be answered.

There is, of course, not just one human genome sequence. The sequence varies between individuals. In 2001, it cost just under $5,300 to sequence a million base pairs of DNA. By April 2013, this cost had dropped to six cents. This means that if you had wanted to have your own genome sequenced in 2001, it would have cost you just over $95 million. Today, you could generate the same sequence for just under $6,000,[6] and at least one company is claiming that the era of the $1,000 genome is here.[7] Because the cost of sequencing has decreased so dramatically, it's now much easier for scientists to study the extent of variation between individual humans, which has led to a number of benefits. Researchers are now able to identify rare mutations that cause severe diseases but only occur in a small number of patients, often in genetically isolated populations such as the Amish communities in the United States.[8] It's possible to sequence tumour cells from patients to identify mutations that are driving the progression of a cancer. In some cases, this results in patients receiving specific therapies that are tailored for their cancer.[9] Studies of human evolution and human migration have been greatly enhanced by analysing DNA sequences.[10]

Honey, I lost the genes

But all this was for the future. In 2001, amidst all the hoopla, scientists were poring over the data from the human genome sequence and pondering a simple question: where on earth were all the genes? Where were all the sequences to code for the proteins that carry out the functions of cells and individuals? No other species is as complex as humans. No other species builds cities, creates art, grows crops or plays ping-pong. We may argue philosophically about whether any of this makes us 'better' than other species. But the very fact that we can have this argument is indicative of our undoubtedly greater complexity than any other species on earth.

What is the molecular explanation for our complexity and sophistication as organisms? There was a reasonable degree of consensus that the explanation would lie in our genes. Humans were expected to possess a greater number of protein-coding genes than simpler organisms such as worms, flies or rabbits.

By the time the draft human genome sequence was released, scientists had completed the sequencing of a number of other organisms. They had focused on ones with smaller and simpler genomes than humans, and by 2001 had sequenced hundreds of viruses, tens of bacteria, two simple animal species, one fungus and one plant. Researchers had used data from these species to estimate how many genes would be found in the human genome, along with data from a variety of other experimental approaches. Estimates ranged from 30,000 to 120,000, revealing a considerable degree of uncertainty. A figure of about 100,000 was frequently bandied about in the popular press, even though this had not been intended as a definitive estimate. A value in the region of 40,000 was probably considered reasonable by most researchers.

But when the draft human sequence was released in February 2001, researchers couldn't find 40,000 protein-coding genes, let alone 100,000. The scientists from Celera Genomics identified 26,000 protein-coding genes, and tentatively identified an additional 12,000. The scientists from the public consortium identified 22,000 and predicted there would be a total of 31,000 in total. In the years since the publication of the draft sequence, the number has consistently decreased and it is now generally accepted that the human genome contains about 20,000 protein-coding genes.[11]

It might seem odd that scientists didn't immediately agree on the numbers of genes as soon as the draft sequence was released. But that's because identifying genes relies on analysing sequence data and isn't as easy as it sounds. It's not as if genes are colour-coded, or use a different set of genetic letters from the other parts of the genome. To identify a protein-coding gene, you have to analyse specific features such as sequences that can code for a stretch of amino acids.

As we saw in Chapter 2, protein-coding genes aren't formed from one continuous sequence of DNA. They are constructed in a modular fashion, with protein-coding regions interrupted by stretches of junk. In general, human genes are much longer than the genes in fruit flies or the microscopic worm called *C. elegans*, which are very common model systems in genetic studies. But human proteins are usually about the same size as the equivalent proteins in the fly or the worm. It's the junk interruptions in the human genes that are very big, not the bits that code for protein. In humans, these intervening sequences are often ten times as long as in simpler organisms, and some can be tens of thousands of base pairs in length.

This creates a big signal-to-noise problem when analysing genes in human sequences. Even within one gene there's just a small region that codes for protein, embedded in a huge stretch of junk.

So, back to the original problem. Why are humans such complicated organisms, if our protein-coding genes are similar to those from flies and worms? Some of the explanation lies in the splicing that we saw in Chapter 2. Human cells are able to generate a greater variety of protein variants from one gene than simpler organisms. Over 60 per cent of human genes generate multiple splicing variants. Look again at Figure 2.5 (*page 18*). A human cell could produce the proteins DEPARTING, DEPART, DEAR, DART, EAT and PARTING. It might produce these proteins in different ratios in different tissues. For example, DEPARTING, DEAR and EAT could all be produced at high levels in the brain, but the kidney might only express DEPARTING and DART. And the kidney cells might produce 20 times as much of DART as of DEPARTING. In lower organisms, cells may only be able to produce DEPARTING and PARTING, and they may produce them at relatively fixed ratios in different cells. This splicing flexibility allows human cells to produce a much greater diversity of protein molecules than lower organisms.

The scientists analysing the human genome had speculated that there might be protein-coding genes that are specific to humans, which could account for our increased complexity. But this doesn't seem to be the case. There are nearly 1,300 gene families in the human genome. Almost all of these gene families occur through all branches of the kingdom of life, from the simplest organisms upwards. There is a subset of about 100 families that are specific to animals with backbones but even these were generated very early in vertebrate evolution. These vertebrate-specific gene families tend to be involved in complex processes such as the parts of the immune system that remember an infection; sophisticated brain connections; blood clotting; signalling between cells.

It's a little as if our protein-coding genome has been built from a giant LEGO kit. Most LEGO kits, especially the large starter boxes, contain a selection of bricks that are variations on a small number of themes. Rectangles and squares, some sloping pieces, perhaps a few arches. Various colours, proportions and thicknesses, but all basically similar. And from these you can build pretty much all basic structures, from a two-brick step to an entire housing development. It's only when you need to build something extremely specialist, like the Death Star, that it's necessary to have very unusual pieces that don't fit the basic LEGO templates.

Throughout evolution, genomes have developed by building out from a standard set of LEGO templates, and only very rarely have they created something completely new. So we can't explain human complexity by claiming we have lots of unusual human-specific protein-coding genes. We simply don't.

But where this all becomes odd is when we compare the size of the human genome with that of other organisms. Looking at Figure 3.1, we can see that the human genome is much bigger than

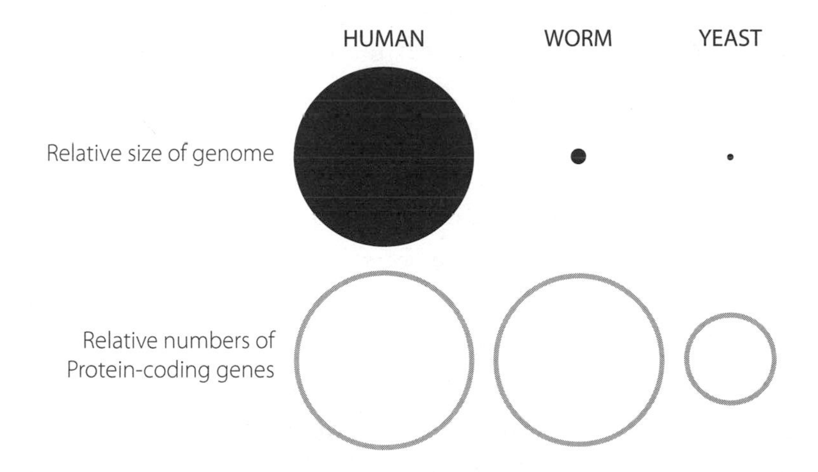

Figure 3.1 In the upper panel, the areas of the circle represent the relative sizes of the genomes in humans, a microscopic worm and single-celled yeast. The human genome is much bigger than those from the simpler organisms. The lower panel represents the relative numbers of protein-coding genes in each of the three species. The disparity here between humans and the other two organisms is much less than in the top panel. The large relative size of the human genome clearly can't be explained solely in terms of numbers of protein-coding genes.

that of *C. elegans* and much, much bigger than that of yeast. But in terms of numbers of protein-coding genes, there isn't anything like as great a difference.

These data demonstrated convincingly that the human genome contains an extraordinary amount of DNA that doesn't code for proteins. Ninety-eight per cent of our genetic material doesn't act as the template for those all-important molecules believed to carry out the key functions of a cell or an organism. Why do we have so much junk?

Poisonous fish and genetic insulation

One possibility is that the question is irrelevant or inappropriate. Maybe the junk has no function or biological significance. It can be a mistake to assume that because something is present, it has a reason to be there. The human appendix serves no useful purpose; it's just an evolutionary hangover from our ancestral lineages. Some scientists speculated back in 2001 that this might also be true of most of the junk DNA in the human genome.

Part of the rationale for this suggestion lay in an interesting animal, the pufferfish (also known as the blowfish). Pufferfish are remarkable creatures. Because they are slow, clumsy swimmers they are unable to evade predators. If faced with a threat, they rapidly take in huge amounts of water and swell up into a globe, which in some species is covered in spikes. If that isn't enough to deter a hungry predator, they also contain a toxin which is over a thousand times more powerful than cyanide. This has given the pufferfish a weird notoriety. In Japan it is considered a delicacy (called fugu), but one with a highly chequered history, since inexpert preparation can carry lethal consequences for the diner.

Genetics researchers were very fond of pufferfish, or at least its DNA. The genome of a particular pufferfish called *Fugu rubripes* is the most compact of any vertebrate. It is only about 13 per cent of the length of the human sequence, but it contains pretty much all the usual vertebrate genes.[12] The reason the pufferfish genome is so small is because it doesn't contain very much junk DNA. In the days when it cost a lot of money to sequence DNA, pufferfish was a very useful species to use when comparing genomes from different organisms. And because its genome contains so little junk, it was relatively easy to identify individual genes, because there weren't the signal-to-noise issues that were such a problem when annotating the human genome. Scientists were able to spot genes in *Fugu rubripes* very easily, and then use the sequence data to help them search for similar genes in noisier genomes such as our own.

Because pufferfish have very little junk DNA but are functional and successful organisms, it was suggested that the non-coding regions of the human genome might be 'simply parasitic, selfish DNA elements that use the genome as a convenient host'.[13] But this isn't necessarily a logical projection. Just because something has no apparent function in a specific organism, it doesn't mean it is irrelevant in all species. Because evolution is usually building from a relatively limited repertoire of components (remember the LEGO set), there is a tendency for features to be co-opted for new functions. So, junk DNA could easily have roles in other organisms, especially ones that are more complex.

It is also worth bearing in mind that there is a functional cost for a cell in containing so much junk DNA. Humans all start life as one cell, formed when an egg fuses with a sperm. That single starting cell divides to form two cells. The two cells divide to form four, and the process continues. An adult human is composed of about 50–70 trillion cells. That's a lot of cells to visualise, so try it this way. If each cell was a dollar bill, and we stacked 50 trillion dollar bills on top of each other, they would stretch from the Earth to the moon and halfway home again.

It takes about 46 cycles of cell division, at a minimum, to create that many cells. And every time a cell divides, it first has to copy all its DNA. If less than 2 per cent of the DNA is important, why would evolution maintain the other 98 per cent if it is simply functionless junk? As we have already acknowledged, the greatest evidence in favour of evolution of species lies in all those things we are stuck with because of our forebears (such as the appendix). But using huge amounts of resources to reproduce 49 'useless' base pairs for every one that performs a function seems like taking redundancy a bit far.

One of the first theories for why the human genome contains so much DNA arose even before the draft human genome sequence had been completed, when researchers already recognised that there was a significant part of our genome that didn't code for protein. It's the insulation theory.

Imagine you own a watch. Not just any old watch, but a phenomenally expensive watch such as a vintage Patek Philippe of the type that sells for a couple of million dollars. Now imagine there is a large and very angry baboon in the vicinity, carrying a really heavy stick. You have to put your watch in a room and you are given a choice. You can't stop the baboon going into any of the rooms, but you can decide on the room where you want to leave the watch. The choices are:

A. A small room with nothing else in it but a table, on which you have to leave the watch.
B. A large room containing 50 rolls of loft insulation, each roll being 5m in length and 20cm deep, and you can hide the watch deep in any one of the 50 rolls.

It's not that difficult to work out which to choose to maximise the chances of the watch escaping damage, is it? And the insulation theory of junk DNA was built on the same premise. The genes that code for proteins are incredibly important. They have been subjected to high levels of evolutionary pressure, so that in any given organism, the individual protein sequence is usually as good as it's likely to get. A mutation in DNA – a change in a base pair – that changes the protein sequence is unlikely to make a protein more effective. It's more likely that a mutation will interfere with a protein's function or activity in a way that has negative consequences.

The problem is that our genome is constantly bombarded by potentially damaging stimuli in our environment. We sometimes think of this as a modern phenomenon, especially when we consider radiation from disasters such as those at the Chernobyl or Fukushima nuclear plants. But in reality this has been an issue throughout human existence. From ultraviolet radiation in sunlight to carcinogens in food, or emission of radon gas from granite rocks, we have always been assailed by potential threats to our genomic integrity. Sometimes these don't matter that much. If ultraviolet radiation causes a mutation in a skin cell, and the mutation results in the death of that cell, it's not a big deal. We have lots of skin cells; they die and are replaced all the time, and the loss of one extra is not a problem.

But if the mutation causes a cell to survive better than its neighbours, that's a step towards the development of potential cancer, and the consequences of that can be a very big deal indeed. For example, over 75,000 new cases of melanoma are diagnosed every year in the United States, and there are nearly 10,000 deaths per year from the condition.[14] Excessive exposure to ultraviolet radiation is a major risk factor. In evolutionary terms, mutations would be even worse if they occurred in eggs or sperm, as they may be passed on to offspring.

If we think of our genome as constantly under assault, the insulation theory of junk DNA has definite attractions. If only one in 50 of our base pairs is important for protein sequence because the other 49 base pairs are simply junk, then there's only a one in 50 chance that a damaging stimulus that hits a DNA molecule will actually strike an important region.

It's also consistent with why the human genome contains so much junk DNA compared with the relatively tiny amounts present in less complex species such as the worm and yeast, as we saw in Figure 3.1. Worms and yeast have short life cycles, and can produce large numbers of offspring. The cost–benefit equation for them is different from that of a species such as humans, who take a long time to reproduce and only have small numbers of offspring. For worms and yeast there probably isn't much point putting a large amount of effort into protecting the protein-coding genes so extensively. Even if a few of their offspring carry mutations that make them less fit for their environment, the majority will probably be OK. But if you get very few shots at passing your genetic material on to the next generation, protecting those important protein-coding genes makes good evolutionary sense.

Nature, as we have seen, is nothing if not adaptive, and so even though the insulation theory makes good sense, it raises another couple of questions. Is insulation the only role of junk DNA?; and where did all this insulating material come from in the first place?

4. Outstaying an Invitation

Every British schoolchild knows the date 1066. It's the year that William the Conqueror and his troops from Normandy in what is modern-day France invaded England. This wasn't some temporary raiding party. The invaders stayed, brought their families over and expanded in numbers and influence. They ultimately assimilated, becoming an integrated part of the English political, cultural, social and linguistic landscape.

Every American schoolchild knows the date 1620. It's the year that the *Mayflower* anchored at Cape Cod, triggering the great wave of European migration and settlement to North America. Like the Normans in Britain over 500 years before them, these early settlers expanded in numbers rapidly, altering the landscape forever.

A similar event happened in the human genome many millennia ago. It was invaded by foreign DNA elements, which then multiplied hugely in number, finally becoming stable integral parts of our genetic heritage. These foreign elements act as a kind of fossil record in our genome, which can be compared with the records from other species. But they also can affect the function of our protein-coding genes, influencing health and disease.

Although they can affect expression of protein-coding genes, these foreign elements don't code for proteins themselves. This makes them an example of junk DNA.

When the draft human genome sequence was released, it was astonishing to realise just how widely these genetic interlopers have spread through our DNA.[1] Over 40 per cent of the human genome is composed of these parasitic elements. They are called interspersed repetitive elements, and there are four main classes.* As their name suggests, they are DNA stretches in which particular sequences are repeated. The sheer numbers are extraordinary. There are over 4 million of these interspersed repetitive elements in the human genome. One class alone is present 850,000 times throughout the genome and constitutes over 20 per cent of our DNA.

Most of these sequences found ways in the past of increasing their numbers within the genome. Often they mimicked the action of certain types of viruses, similar to the virus that causes AIDS. The basics of this are shown in Figure 4.1. It provides a mechanism whereby a cellular sequence can be copied over and over again

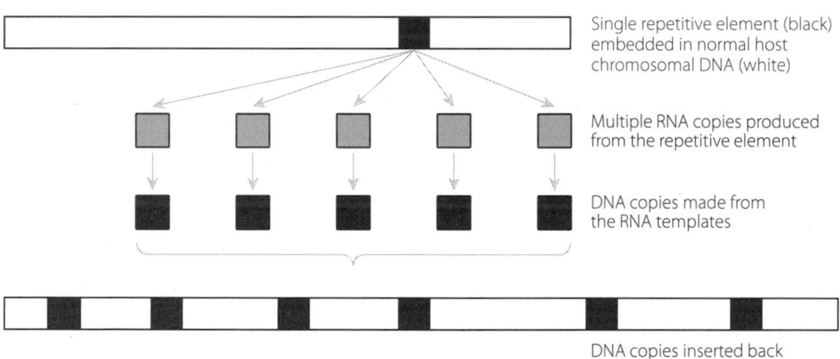

Figure 4.1 A single DNA element is copied to create multiple RNA copies. In a relatively unusual process, these multiple RNA molecules can be copied back into DNA and reinserted into the genome. This amplifies the number of these elements. This may have happened multiple times in early evolution, but just one round is shown here for clarity.

* The four classes are known as SINES (short interspersed repetitive repeats); LINEs (long interspersed elements); LTRs (elements with long terminal repeats); DNA transposons.

and reinserted back into the genome. This creates an amplification cycle that results in the repetitive sequences increasing in number faster than the rest of the genome.

In many ways, the repeats have undergone the equivalent of copy-and-paste in the genome. This is what has allowed them to spread all over our chromosomes.

As a consequence of these amplifications, we carry enormous numbers of these elements in our genome. The question is whether or not this actually matters. Do these sequences have any effect, or are they just passengers in the genome, with neither positive nor negative impacts?

There are various ways in which we can consider this question. Most of the repeats are very old in evolutionary terms. Comparisons with other animals show that the majority of the repeats arose before placental mammals separated from other animal lineages, over 125 million years ago. For at least one of the classes of repeats, we haven't developed any new insertions since we separated from the Old World monkeys about 25 million years ago. So there seems to have been a huge expansion in repeats in the human genome in our distant past. After that, the numbers didn't increase significantly, which might suggest that there is an upper limit to the number of these repeats we can tolerate. But they also seem to be cleared out of the genome very slowly, which in turn suggests that as long as the number of repeats is below this limit, we can put up with them.

And yet there does seem to be some difference in the ways that the human genome copes with such repeats, compared with other species. Mammals in general seem to have a more diverse range of certain repeats than other species. But in mammals, these are based on very ancient sequences that have stuck around for a long time. In other organisms, the old repeats have been cleared out to some extent, and newer ones have taken their places. The authors of the draft human genome sequence calculated that in the fruit fly, a non-functional DNA element has a half-life of about 12 million years. In mammals, the half-life is about 800 million years.

But even among mammals, humans seem to be unusual. Repeat elements have been decreasing in number in the hominid lineage since the expansion in the number of mammalian species. This hasn't happened in rodents. The majority of the repeats in the human genome also no longer undergo copy-and-paste. Essentially, the repeats are more active in rodents than in primates.

Perhaps as a consequence, repeats are a bigger cause of problems in rodents than in humans. If repeats replicate in the genome, they may insert into or near functional protein-coding genes and interfere with their normal roles. In some cases they may prevent the correct protein from being expressed. In others, they may drive increased expression of the protein. In mice, insertion of repeats into novel regions of the genome is 60 times more likely to be the cause of a new genetic condition than is the case in human cells. In mice, these account for 10 per cent of all new genetic mutations, whereas the figure is one in 600 for humans. We seem to have our genomes under tighter control than our rodent cousins.

Dangerous repetition

Perhaps this is just as well, when we look at some of the consequences of this kind of mutation mechanism in rodents. There's a mouse strain in which such a mutation results in no tail. This in itself might not be too problematical, but the kidneys also fail to develop, and that's a very bad thing indeed.[2] This is because the insertion leads to over-expression of a nearby gene. In a different strain, the insertion switches off an important gene in the central nervous system. This results in mice that spasm if they are handled, and have a lifespan of just two weeks.[3]

We can also draw a similar conclusion about the potential

impact of such repeats from the opposite phenomenon, i.e. by looking at regions of the genome where these repeats hardly ever occur.

There is group of genes called the HOX cluster, which is very important in driving the correct development of complex cellular organisms. The genes in the cluster are switched on in a specific order during development, and expressed at highly regulated levels. If anything goes wrong with this order, the effects can be very profound. The importance of the HOX cluster was first shown in fruit flies. Flies with mutations in these genes developed some extraordinary characteristics. In the most famous example, the flies didn't have antennae on their heads. Instead, their heads had a pair of legs on them.[4]

Just like flies, mammals also rely on the appropriate expression patterns of HOX genes for the development of the correct body patterns. Mutations at the HOX cluster are rare in humans, probably because these genes are so important. But it has been shown that a mutation in at least one HOX gene results in defects in the ends of the limbs.[5]

The HOX cluster is one of the few places in the human genome that is almost completely clear of interspersed repetitive elements. This suggests that even relatively benign genetic interlopers have the potential to affect gene expression, and that there are some regions of the genome where evolution has ensured that they are kept at bay. This repeat-free aspect of the HOX cluster is also found in other primates and in rodents.

The presence of interspersed repeats in the genome can have unexpected consequences. There's an unusual class of repeats caused ERVs. ERV stands for endogenous retrovirus. The human immunodeficiency virus (HIV, the causative agent of AIDS) is an example of a retrovirus. Such viruses are characterised by the genetic material being made of RNA, not DNA. The viral RNA is copied to form DNA, which can then integrate into the host genome. The host treats the DNA like its own, producing new viral components and ultimately new viruses.

Long ago in our evolutionary history, some retroviruses became fully established in our genomes. Many are now genomic fossils. Certain parts of the retroviral sequences have been lost, and so they can never again produce viral particles. But some still contain all the components required to make new viruses. These are normally kept under very tight control by the cell.[6] Scientists have also discovered that the immune system doesn't just fight off viruses that infect us from the outside world; it also plays a role in keeping these endogenous viruses under control. Genetically engineered mice which lack certain components of the normal immune system suffer problems through the reactivation of these viruses lurking in their own genomes.[7]

This control of endogenous retroviruses is a potential issue in one approach to tackling a problematic area of human health. Every year, thousands of people die on waiting lists for organ transplants because there aren't enough donors. For example, approximately one in three of the people whose lives could potentially be saved by a heart transplant dies while still on the waiting list.[8]

One potential way around this would be if we could use hearts from animals as replacement organs. This is known as xenotransplantation ('xeno' is derived from the Greek for 'foreign'). For cardiac transplants, the animal of choice is the pig. Its heart is about the same size and strength as the human organ.

There are a number of technical hurdles to overcome (in addition to ethical issues around the use of pigs that matter to certain religious groups).[9] Some of these are being addressed by the creation of genetically modified pigs that don't provoke the very aggressive immune response that is a problem when introducing pig cells into the human cardiovascular system. But there may be another issue. The pig genome contains endogenous retroviruses, just as the human genome does. But the ones in pigs are different from the ones in humans. Work at the end of the 20th century showed that some of these pig retroviruses can infect human cells, given the right conditions.[10]

There's a possible scenario that has worried some scientists. Anyone who receives a pig heart will inevitably be receiving immunosuppressive drugs to prevent rejection of the foreign organ. Reactivation of endogenous retroviruses is more likely when individuals are immunosuppressed. Human systems have evolved in part to control the endogenous retroviruses that have been in our genome since we evolved. But they may not be as efficient at controlling the ones hiding in the pig genome. This theoretically could mean that the endogenous retroviruses could escape from the pig heart and attack and enter other cells in the human recipient. From there, they might even escape into the wider population.

More recent data have suggested that the risk of this happening has perhaps been overstated in the past,[11] but it's certainly an area of junk DNA that will require close scrutiny if xenotransplantation is to become a reality.

Other repeated sequences in the genome can cause health problems more directly. There are some parts of the genome where large sections, sometimes hundreds of thousands of base pairs in length, were duplicated relatively recently during human evolution. The 'original' and the 'duplicate' may end up in very different parts of the genome, even on different chromosomes from one another.

These regions can cause problems when eggs or sperm are being formed. During this formation, there is a very important stage where chromosomes undergo a process called crossing-over. A chromosome inherited from your mother pairs up with the equivalent chromosome inherited from your father, and they swap bits of DNA between the two. It's a way of increasing the amount of variation in the gene pool, by mixing up combinations of genes. If there are two parts of the genome that look very similar because of repeat sequences but which are not actually a matching pair of chromosomes, this crossing-over may occur between regions of the genome that aren't meant to swap material. The consequence may be that eggs or sperm are produced that have extra sections of DNA, or are missing critical regions.[12]

This can lead to disease in individuals who inherit these genomic defects. One example is Charcot-Marie-Tooth disease, where there are defects in the nerves that transmit sensation and control motor functions.[13] Another is Williams-Beuren syndrome, a condition characterised by developmental delay, relative shortness, a range of unusual behavioural traits combined with mild learning disability, and long-sightedness.[14]

The duplicated regions in the genome that give rise to the problems during crossing-over often contain multiple protein-coding genes. It's probably not surprising that the symptoms in patients affected by abnormal crossing-over are often quite complex. It's likely that more than one pathway is affected by the change in the number of multiple genes.

It might seem odd that these duplicated regions have been retained during human evolution, if they can give rise to such problems. But in reality, most of the time the cells that form eggs and sperm perform crossing-over really well, and don't mix up the wrong parts of chromosomes. The duplications have also acted as a way that the human genome has been able to increase the numbers of certain genes quite rapidly, in evolutionary terms. This can be useful. The 'spare' copy may act as the raw material for evolutionary adaptation. A few changes to the protein-coding gene sequence can create a protein with a related but discrete function from the original. This may be how the large family of genes that allows mammals to detect a huge range of different smells evolved.[15] It's another example of the parsimony with which the human genome has evolved, adapting existing genes and proteins, rather than starting from scratch. A genomic two-for-one offer.

From guilt to innocence via junk DNA

Most of the junk repetitive DNA that we have considered so far in this chapter is formed of quite large units. These tend to be at least 100 base pairs in length and are frequently much longer. That's partly why they account for so much of the genome. But there are other junk repetitive units that are much smaller, based on repeats of just a few base pairs. These are called simple sequence repeats. We already met a few examples of these in the exploration of Fragile X syndrome, Friedreich's ataxia and myotonic dystrophy. In each of these cases, three-base-pair sequences were repeated a number of times, and reached their maximum in patients with the disorders.

Repeats of short motifs account for about 3 per cent of the human genome. They are very variable between individuals. Let's consider an arbitrary repeat of two base pairs, say GT, at a particular position on chromosome 6. I may have inherited eight copies (sequence would be GTGTGTGTGTGTGTGT) on chromosome 6 from my mother and seven copies on chromosome 6 from my father. You, on the other hand, may have inherited ten copies from your mother and four from your father.

These simple sequence repeats have proved to have great usefulness because they are found all over the genome, vary a lot between individuals at each position where they occur in the genome and are easy to detect using cheap, sensitive methods.

Because of these characteristics, such repeats are now used for DNA fingerprinting. This is the process by which blood or tissue samples can be unequivocally associated with a specific individual. This has facilitated paternity testing and revolutionised forensic science. Its applications in the latter have included identification of victims of massacres, convictions of the guilty and exonerations of the innocent, including cases where the wrong person has been in jail for decades. Over 300 people in the United States have been freed after DNA testing established their innocence, nearly 20 of whom had been on death row at some point during their incarceration.[16] Additionally, in about half of these cases, DNA evidence was able to determine the real guilty party.

Not bad for a bit of junk.

5. Everything Shrinks When We Get Old

The movie *Trading Places*, starring Dan Aykroyd, Eddie Murphy and Jamie Lee Curtis, was a huge hit in 1983, grossing over $90 million at the US box office.[1] It's a convoluted comedy but the premise behind it is the exploration of genes versus environment. Is a successful man successful because of intrinsic merit or because of the environment in which he is placed? The movie comes out firmly on the side of the latter.

A similar phenomenon can happen in our genomes. An individual gene may perform a relatively innocuous role, helping a cell keep on keeping on, so to speak. The gene produces protein at just the right rate to do this job. A major factor in controlling the amount of protein that is produced is the position of the gene on the chromosome.

Now let's imagine that the gene is transported to a new neighbourhood, like Dan Aykroyd's character ending up in the slums or Eddie Murphy's character finding himself transported to a mansion. In this neighbourhood, our transported gene is surrounded by new genomic information, which instructs it to make much higher amounts of protein. The high levels of the protein whip the cell forwards, pushing it to grow and divide much faster than usual. This can be one of the steps that leads to cancer. There's nothing bad about the gene itself, it's just in the wrong place at the wrong time.

This process is caused when two chromosomes break in a cell at the same time. When a chromosome breaks, a repair machinery immediately targets the break and joins the two bits up again. This is usually a pretty slick process. But if two (or more) chromosomes break at the same time, there can be problems. The ends of the chromosomes may become joined up incorrectly, as shown in Figure 5.1. This is how a 'good' gene may end up in a 'bad' neighbourhood, and begin causing problems. This is particularly an issue because the rearranged chromosomes will be passed on to all daughter cells every time cell division takes place. Probably the most famous example of this mechanism is in a human blood cancer called Burkitt's lymphoma, where there is a rearrangement between chromosomes 8 and 14. This results in very strong over-expression of a gene* that encourages cells to proliferate aggressively.[2]

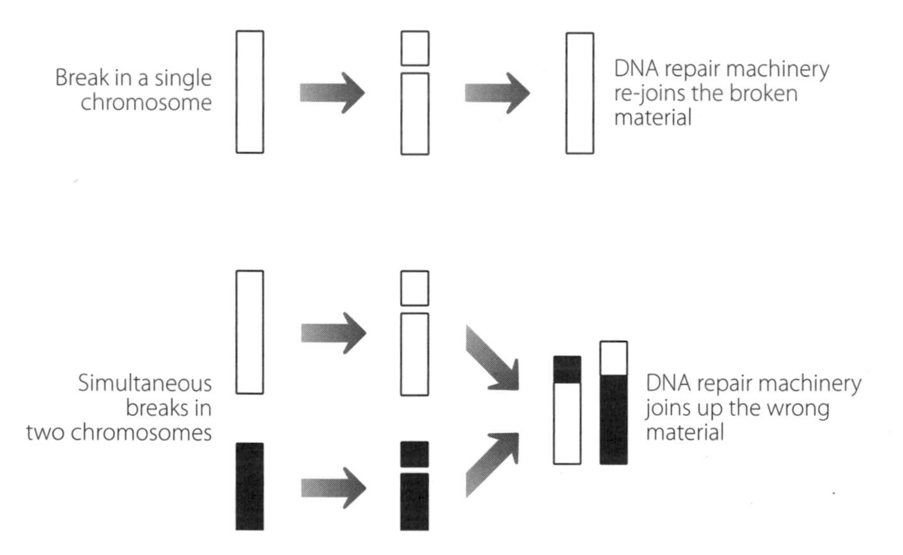

Figure 5.1 In the upper panel a single chromosome breaks and is repaired by the cell. In the lower panel two chromosomes break simultaneously. The cell machinery may be unable to work out which break occurred on which chromosome. The chromosomes may be joined together inappropriately, creating hybrid structures.

* The gene is called Myc.

Luckily, it's probably quite rare that two chromosomes break at exactly the same time. More frequently there will be a time difference. So, the machinery that repairs DNA has evolved to act really quickly. After all, the faster it repairs a break, the lower the chance that there will be multiple breaks present at the same time in an individual cell. The DNA repair machinery starts to operate as soon as the cell detects that there is a broken piece of DNA. It does this by having mechanisms to detect the end of the break.

But this creates a whole new set of problems. Our cells contain 46 chromosomes, each of which is linear. In other words, our cells always have 92 chromosome ends, one at each end of a chromosome. The DNA damage machinery has to have a way of distinguishing the perfectly normal ends of chromosomes from the abnormal ends caused by breakages.

DNA shoelaces

The way that cells have solved this is to have special structures on the normal ends of the chromosomes. Are you wearing shoes with laces? If so, have a quick look at those laces. At either end there is a little cap made from metal or plastic. This is called the aglet, and it stops the lace from unravelling and fraying. Our chromosomes have their own aglets, and these are extremely important for maintaining the integrity of our genome.

These chromosomal aglets are called telomeres and they are made from a form of junk DNA that we have known about for many years, plus complexes of various proteins. The telomeric DNA is formed from repeats of the same six base pairs, TTAGGG, repeated over and over again.[3] These stretch for an average of about 10,000 base pairs in total on each end of every chromosome in the umbilical cord blood of a newborn human baby.[4]

The telomeric DNA is bound by complexes of proteins that help to maintain the structural integrity.* The term telomere really refers to the combination of the junk DNA and its associated proteins. A graphic demonstration of the importance of these proteins was shown by some researchers working in mice in 2007. They knocked out expression of one of the proteins by completely inactivating its gene, and found that the resulting mice embryos died early in development.**

When the researchers examined the chromosomes in these genetically modified mice, they found that many of them had joined up. The ends had linked up with each other. This was because the DNA repair machinery no longer recognised the telomeres as telomeres. Instead, it reacted as if faced by a whole slew of broken chromosomes and did what it does best. It stuck them together. Unfortunately, by doing so, gene expression became completely disordered. Eventually the chromosomes and cells became so dysfunctional that they triggered a type of cellular suicide,*** halting development completely.

There is also another feature of the telomeres that is of major interest in biology and human health. Back in the 1960s, researchers were studying how cells divide in the laboratory. They didn't work with cancer cell lines, as these are derived from cells that have become immortal through abnormal changes. Instead, they studied a kind of cell known as a fibroblast. Fibroblasts are found in a wide range of human tissues. They secrete something called the extracellular matrix, a sort of thick wallpaper paste that holds

* Yes, I do like Star Trek. Occasionally.

** The gene was called *Gcn5*. It codes for a protein with a number of functions, one of which is to add a small molecular group called acetyl to the amino acid lysine in proteins.

*** The technical terms for this cell suicide are programmed cell death, or apoptosis.

the cells in position. It's relatively easy to take a biopsy, for example from skin, and isolate the fibroblasts. These will grow and divide in culture. What the researchers discovered all those years ago was that the cells wouldn't keep dividing forever. There came a point when they stopped dividing, even when supplied with all the nutrients and oxygen they needed. The cells didn't die, they just stopped proliferating. This is known as senescence.[5]

Scientists later realised that the telomeres in cells became shorter with each cell division. Every time one of the cells divided, all the DNA in that cell was copied. This ensured that both daughter cells inherited the same 46 chromosomes as the mother cell. But the system that copies the DNA in chromosomes can't get right to the ends. So, over progressive cycles of cell division, the telomeres became shorter and shorter.[6]

But this didn't prove that the shortening of the telomeres actually caused cell senescence. It was perfectly possible that the effect on telomere length acted as a kind of marker for cell proliferation, but didn't have any actual role to play in the changes in cell behaviour.

This is a really important concept in scientific enquiry. There are plenty of situations in which we can see a correlation between two things, but we shouldn't from that automatically assume there is a causal relationship. Consider the following relationship. There is a strong relationship between developing lung cancer and sucking cough sweets. This doesn't of course prove that sucking cough sweets gives you lung cancer. One of the first symptoms of lung cancer in many people is the development of a persistent cough, and someone with a cough is likely to try sucking hard sweets to decrease their discomfort.

The confirmation that telomere shortening did indeed lead to senescence came in the 1990s. Scientists demonstrated that if they increased the length of the telomeres in fibroblasts, the cells would bypass senescence and grow indefinitely.[7]

It is now generally accepted that the telomeres act as a molecular clock, counting us down as we age. Not all the details have been established yet, because it's a difficult area of biology to investigate, for a variety of reasons. One is that in any given cell, the 92 telomeric regions (one at each end of each chromosome) won't be the same length. This makes it hard to come up with a meaningful measure of telomere length that is applicable throughout a cell, never mind an entire human being.[8] It's also very difficult for scientists to use their favourite model animal – the mouse – to investigate the relationships between telomere biology and ageing. This is because rodents have extremely long telomeres, much lengthier than in humans. Rodents, of course, are much shorter-lived than humans, suggesting that telomere length is not the only arbiter of ageing, but the accumulated evidence suggests that in humans they are of major importance.

Looking after the shoelaces

What we do know is that our cells don't succumb to the ageing process without a fight. They contain mechanisms to try to keep the telomeres long and intact as much as possible. This is achieved in our cells by something called telomerase activity. The telomerase system adds new TTAGGG motifs onto the ends of the chromosomes, basically restoring these important bits of junk DNA that are lost when the cells divide. Telomerase activity requires two components. One part is an enzyme, which adds the repeated sequences back on to the chromosome termini. The other is a piece of RNA, of a defined sequence, which acts as a template so that the enzyme adds the correct bases.

So the ends of our chromosomes rely heavily on junk DNA, genomic material that doesn't code for proteins. The telomeres themselves are junk, and to maintain them the cell uses the output from a gene that produces RNA, but which is never used as a template for a protein. This RNA itself is a functional molecule, carrying out a vital role.*[9]

But if our cells contain a mechanism for maintaining telomere length, through the activity of the telomerase system, why do the telomeres get progressively shorter? What's wrong with the system, why doesn't it work properly?

The reason probably stems from the fact that there are few systems in biology that work well if allowed to run unchecked. And telomerase activity is held in very tight check indeed in our cells. The pathological exception to this is in cancer cells. Cancer cells frequently have adapted in such a way that they express high levels of telomerase activity and have elongated telomeres. This contributes to the aggressive growth and proliferation of many tumours. Our cellular systems have probably reached an evolutionary compromise. The telomeres are maintained at sufficient levels that we live long enough to reproduce (anything after that is irrelevant in evolutionary terms). But they aren't so long that we succumb to cancer too early.

The basic telomere length in an individual is set fairly early in development, at a time when there is an uncharacteristic spike in the telomerase activity.[10] Telomerase activity is also high in germ cells, the cells that give rise to eggs and sperm.[11] This is to ensure that our offspring inherit telomeres of a good length.

Many human tissues contain cells known as stem cells. These are responsible for producing replacement cells when needed. When new cells are needed, a stem cell will copy its DNA and then split it between two daughter cells. Typically, one of these daughter cells will develop into a fully fledged replacement cell. The other will become a new stem cell, which can continue to create replacements in the same way.

---

* The core enzyme is encoded by the TERT gene and the RNA template is encoded by the TR gene, also known as TERC.

One of the 'busiest' cell types in the human body is the type of stem cell that gives rise to all the blood cells,* including red blood cells and those that we rely on to fight infection. These stem cells proliferate at an incredible rate. This is because we constantly need to replenish the immune cells that fight off the foreign pathogens we encounter every day of our lives. We also need to replace red blood cells, because these only survive for about four months. Incredibly, the human body produces about 2 million red blood cells every second.[12] That requires an awfully active stem cell population, in a pretty much constant state of cell division. These stem cells are enriched for telomerase activity, but eventually even they suffer from telomeres that are too short to do their job properly.[13,14] This is one reason why the elderly are at greater risk of infection than younger adults. They are essentially running out of immune cells. It's also one of the reasons why cancer rates rise with age. Our immune system usually does a good job of destroying abnormal cells, but the effectiveness of this surveillance declines as stem cells die off.

Why is the length of our telomeres so important? It's only junk DNA, so why should it matter if there are only several hundred copies of the non-coding TTAGGG, rather than a few thousand? Much of the problem seems to lie in the relationship between the DNA at the telomeres and the protein complexes that are deposited on this DNA. If the repetitive DNA shrinks below a critical level, the end of the chromosome can't bind enough of the protective proteins. We've already seen one of the consequences of a lack of the relevant proteins in the mice that died before birth.

That was a very extreme example, but it's undoubtedly the case that it's vital that the telomeres are long enough to bind lots of the protective protein complexes. We know that this is true in humans as well as mice, because there are people who have

* The technical name for this population is the haematopoietic stem cell (HSC).

inherited mutations in certain key components of the systems for maintaining the telomeres. The effects witnessed aren't as dramatic as in the genetically modified mice, but that's because such severely affected foetuses will tend to be lost during pregnancy. But the mutations we know about lead to conditions associated with certain disorders that are normally age-related.

Telomeres and diseases

The disorders are predominantly caused by mutations in the telomerase gene, or in the gene that codes for the RNA template, or in genes that encode proteins that protect the telomeres, or help the telomerase system to work effectively.*

Essentially, mutations in any of these genes can have similar effects. They basically make it harder for cells to maintain their telomeres. Consequently, the telomeres in patients with these mutations shorten more rapidly than in healthy individuals. This is why they develop symptoms that are suggestive of premature ageing. These disorders are known as human telomere syndromes.[15]

Dyskeratosis congenita is a rare genetic condition, affecting about one in a million individuals. Patients suffer from a whole raft of problems. Their skin contains random dark patches. They develop white patches in their mouth, which can progress to oral cancer, and their fingernails and toenails are thin and weak. They suffer progressive and seemingly irreversible organ failure, triggered initially by bone marrow failure and lung problems. They are also at increased risk of cancer.

Scientists have realised that this condition can be caused by mutations in different genes in different affected families. At least eight mutated genes are known at the moment, and it's quite possible that there are more.[16] The feature that all the genes have in

* This gene is called Dyskeratosis congenita 1 (DKC1) or dyskerin.

common is that they are involved in maintaining telomeres. This shows us that no matter how this region of junk DNA gets messed up, the final symptoms tend to be similar.

The lung problems are known as pulmonary fibrosis. Patients suffering from this condition have debilitating symptoms. They suffer shortness of breath and cough a lot, because they can't move carbon dioxide out of their lungs efficiently or get oxygen into them easily. Looking at their lungs down a microscope, pathologists can see substantial regions where the normal tissue has been replaced by inflammation and fibrous tissue, rather like scar formation.[17]

These clinical and pathological findings in the lungs are ones that are seen quite commonly in respiratory disease, and this prompted scientists to look at samples from patients with a condition known as idiopathic pulmonary fibrosis. Idiopathic just means that there is no obvious reason for the disease. Researchers tested these patients to see if any of them also had defects in the genes whose products protect the telomeres. In all, up to one in six people with a family history of this disease, but no previously identified mutations, were shown to have defects in the relevant genes.[18,19] Even in patients where there was no apparent family history of pulmonary fibrosis, mutations in telomere-relevant genes were found in between 1 and 3 per cent of cases.[20,21] There are about 100,000 patients with idiopathic pulmonary fibrosis in the United States, so at a conservative estimate 15,000 of them probably have developed the disease because they cannot maintain their telomeres properly.

Defects in the mechanisms that protect telomeres can also cause a different disease. There's a condition called aplastic anaemia, in which the bone marrow fails to produce enough blood cells.[22] It's rare, affecting about one person in half a million. About one in twenty of the people with this condition have mutations in the telomerase enzyme or the accessory RNA template.

What may be happening in some of these patients is that they have both bone marrow defects and lung defects, but one problem becomes clinically apparent before the other. This can lead to unexpected consequences when medically treated. Bone marrow transplants are one of the treatments used for patients with aplastic anaemia. The patients are given drugs to prevent their immune system from rejecting the new bone marrow. Some of these drugs are known to have toxic effects in the lungs. For most patients with aplastic anaemia, this isn't really a problem. But for those patients who have defects in their telomerase system, these drugs can trigger lung fibrosis that may actually be lethal.[23] The cure becomes the cause of death.

There's an odd genetic reason why clinicians may not realise that the symptoms they see in a patient are part of an inherited telomere problem. The telomerase complex is usually active in the germ cells, so that parents pass on long telomeres to their children. But in some of the families where there are mutations in the genes encoding the telomerase enzyme or the accessory RNA factor, this isn't the case. As a consequence, each generation passes on shorter telomeres to its offspring. Because symptoms develop when the telomeres fall below a certain length, each successive generation is born rather nearer to the point where their telomere length falls over the cliff edge.[24]

The effects of this are quite dramatic. A grandparent may have relatively long telomeres and develop pulmonary fibrosis in their 60s. Their child may have intermediate-length telomeres and develop lung symptoms in their 40s. But the third generation may inherit really short telomeres. They may develop aplastic anaemia in childhood.

Because the grandparental and parental generations' conditions don't develop until quite late in life, the grandchild may become sick before any of its elders have started displaying symptoms. This will make it difficult for a clinician to recognise that a genetic disease is present in the family, and this is compounded by the different symptoms found in the most severely and least severely affected individuals.

This strange pattern, where the oldest generation has different and milder symptoms that develop later in life than those found in the youngest generation, is rather similar to the inheritance pattern we saw in Chapter 1 for myotonic dystrophy. This is a very unusual genetic phenomenon and it is striking that in the two most clear-cut examples of this, the effect is ultimately caused by a change in length of a stretch of junk DNA.

One obvious question is why some tissues are more susceptible to short telomeres than others. This isn't altogether clear, but some interesting models are emerging. It's likely that tissues where there is a lot of proliferation will be susceptible to defects that lead to shorter telomeres. The classic example is the blood stem cell population, as described earlier in this chapter. If these cells have difficulties maintaining the length of their telomeres then eventually the stem cell population will run out.

That seems like a possible explanation for aplastic anaemia but it won't work for pulmonary fibrosis. Lung tissue replicates quite slowly, yet pulmonary fibrosis is common in people with telomere defects. It's possible that in lung cells the effects of shortened telomeres operate in tandem with other factors that affect the genome and cell function. These take time to develop, so lung symptoms typically develop later than ones that are caused by problems with the blood stem cells.

Our lungs are exposed to potentially damaging chemicals with every breath we take, so perhaps it's not surprising that they struggle to tolerate the burden of defective telomeres. One of the most common sources of dangerous inhaled chemicals is tobacco. The global impact of smoking tobacco on human health is huge. The World Health Organization estimates that nearly 6 million people die every year as a consequence of smoking, over half a million of them from the effects of second-hand smoke.[25]

Researchers examined the effects of cigarette smoke experimentally. They genetically manipulated mice so that some of them had short telomeres and then exposed various mice to cigarette smoke.[26] The results are shown in Figure 5.2. Essentially, the only mice that developed pulmonary fibrosis were those that had short telomeres and were exposed to cigarette smoke.

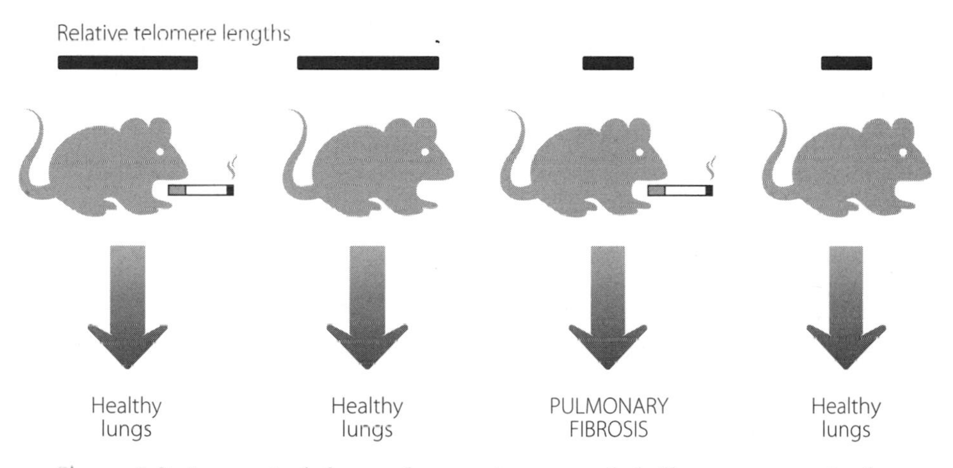

Figure 5.2 A genetic defect and an environmental challenge are required to produce pulmonary fibrosis in mice. Mice with shortened telomeres don't develop fibrosis, and nor do mice exposed to cigarette smoke. But mice with the double insult of shortened telomeres and exposure to cigarette smoke do develop the condition.

Cigarette smoking is not the only factor that affects human health, of course, although not smoking is probably the single smartest thing you can do for yourself. But the major factor that affects human health in wealthy countries is age itself. This wasn't always the case. But it has been true since we made giant medical, pharmacological, social and technological progress in combating what used to kill us early: all those old-fashioned things like infectious diseases, early childhood mortality and malnutrition.

Tick-tock goes the telomere

Getting old is now the major risk factor for development of chronic conditions. That's a big problem when we realise that by 2025 there are likely to be over 1.2 billion people above the age of 60 worldwide.[27] Cancer rates rise dramatically over the age of 40. If you live to 80, there's an even chance you will develop some type of cancer. If you are over 65 and you're an American, there's about the same chance you will have cardiovascular disease.[28] There's plenty more statistics that paint a similarly bleak portrait, but why depress ourselves? Oh what the heck, one last one: the Royal College of Psychiatrists in the UK has stated that about 3 per cent of over-65s have clinical depression and one in six has symptoms of milder depression that are noticed by others.[29]

Yet we all know that two individuals of the same chronological age may be very different in their health. Steve Jobs, the co-founder of Apple, died from cancer at the age of 56. Fauja Singh ran his first marathon at the age of 89, and his last at the age of 101 (no, it wasn't the same one). There's a lot we don't know about what controls longevity – it is almost always a combination of genetics, environment and sheer luck. But what we do know is that simply counting how many years someone has been alive only gives you a very partial picture.

We are starting to realise that telomeres may be quite a sophisticated molecular clock. The rate of telomere shortening can be influenced by environmental factors. This means we may be able to use them as markers not of simple chronology, but of healthy years. The data are rather preliminary and not always consistent. This is partly because measuring telomeres in a consistent way is challenging, as described earlier, and we usually measure them in cells that we can access easily. These are typically the white blood cells, and they may not always be the most relevant cell type to examine. But despite these caveats, some intriguing data are emerging.

Let's go back to our old enemy, tobacco. One study analysed the length of telomeres in the white blood cells of over 1,000 women. They found that the telomeres were shorter in those who smoked, with an increased rate of loss of about 18 per cent for every year of smoking. They calculated that smoking 20 cigarettes a day for 40 years was equivalent to losing almost seven and a half years of telomere life.[30]

A 2003 study looking at mortality rates in the over-60s claimed that the people with the shortest telomeres had the highest mortality rates.[31] This was mainly driven by cardiovascular mortality and the findings have been supported by a later, larger study in a different elderly population.[32] A study in a group of centenarians from the Ashkenazi Jewish community found that longer telomeres were associated with fewer symptoms of the diseases of ageing, and with better cognitive function than that found in people of a similar advanced age but with shorter telomeres.[33]

Sometimes we forget that it's not just physical factors that affect health and longevity. Chronic psychological stress can be very harmful for an individual, with negative impacts on multiple systems including their cardiovascular health and their immune responses.[34] Individuals who suffer chronic psychological stress tend to die younger than less stressed individuals. A study of women aged between 20 and 50 showed that those in the chronically stressed group had shorter telomeres than the unstressed women. This was calculated to equate to about ten years of life.[35]

In the great pantheon of global human health problems that are eminently avoidable but having terrible impact, obesity seems to be on a mission to duke it out with smoking. Turning again to the World Health Organization we learn that nearly 3 million adults die each year because of being obese or overweight. Nearly a quarter of the burden of heart disease is attributable to people being overweight or obese. For type 2 diabetes, the contribution of obesity is even worse (almost half of all cases are caused by being overweight) and it's also true for a significant proportion of cancers (between 7 and 41 per cent).[36] The economic and social costs of this global epidemic are frightening.

Recent data have shown that there is significant interaction between the systems in our cells that try to regulate and respond to energy and metabolism fluctuations, and those that maintain genomic integrity, including telomere stability.[37] It's unsurprising, therefore, that scientists have analysed the lengths of telomeres in cells from obese individuals. The same paper that examined the effects of smoking on telomere length also looked at the effects of obesity. They found that the telomere shortening associated with obesity was even more pronounced than for smoking, equating to nearly nine years of life.[38]

If all this inspires you to keep your weight under control, choose how you do this rather carefully. According to the United Nations, the country with the highest percentage of people who are aged 100 or over is Japan.[39] The traditional Japanese diet almost certainly plays a role in this, because Japanese people who have changed to a Western diet develop Western chronic diseases. The traditional diet is based on low protein intake and relatively high carbohydrate levels. Studies in rats also showed that a low-protein diet early in life was associated with increased lifespan, which in turn was associated with long telomeres.[40]

So if you're thinking of adopting the high-protein and low-carb Atkins or Dukan diets, have a little word with your junk DNA first. I suspect your telomeres might say no.

6. Two Is the Perfect Number

One cell becomes two; two become four; four become eight and, to quote from *The King and I*, 'et cetera, et cetera, and so forth'[1] until there are over 50 trillion cells in a human body. Every time a human cell divides, it has to pass on exactly the same genetic material to both daughter cells as it contains itself. In order to do this, the cell makes a perfect copy of its DNA. This results in a replicate of each chromosome. The two replicates stay attached to each other initially, but then are pulled apart to opposite ends of the cell. A basic schematic for this is shown in Figure 6.1.

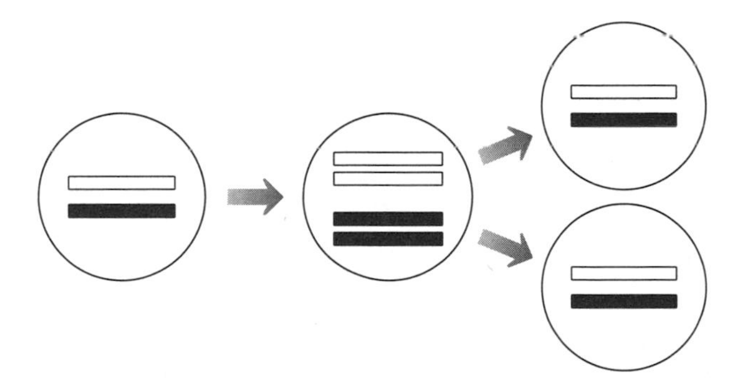

Figure 6.1 A normal cell contains two copies of each chromosome, one inherited from each parent. Before a cell divides, each chromosome is copied to create a perfect duplicate. The copies are pulled apart when the cell divides. This creates two daughter cells, containing exactly the same chromosomes as the original cell. For simplicity, this figure shows just one pair of chromosomes, rather than the 23 pairs in a human cell. The different colours indicate different origins of the pair, one from each parent. The diagram only shows division of the nucleus, but this is also accompanied by division of the rest of the cell.

The only exception to this is when the germ cells in the ovaries or testes create eggs or sperm. Eggs or sperm only contain half the number of chromosomes that are found in all the other cells of the body. The result of this is that when an egg and a sperm fuse, the full chromosome number is restored in the single cell (the zygote) which will then divide to become two cells et cetera, et cetera and so forth.

This halving of the chromosome number is possible because all our chromosomes come in pairs. We inherit one of each pair from our mother and one from our father. Figure 6.2 shows how the chromosome number is halved when eggs or sperm are created.

If cell division goes wrong, either when new body cells are created or when the germ cells create eggs or sperm, the effects can be really serious, as we will see later in this chapter. Cell division is an exceptionally complex process, involving hundreds of different proteins working in a highly coordinated fashion. Given how complicated it is, and how vital it is that cell division happens smoothly and successfully, it might seem surprising that quite a lot of it is critically dependent on a long stretch of junk DNA.

This particular stretch of junk DNA is called the centromere, and unlike the telomeres from the last chapter, the centromere is found on the interior of a chromosome. Depending on the chromosome, it may be pretty much in the middle, or it may be near to an end. Its position is consistent in the sense that on human chromosome 1, for example, it's always near the middle whereas in human chromosome 14 it's always near the end.

Centromeres are essentially attachment points for a set of proteins that drag the separated chromosomes to opposite ends of the cell. Imagine Spider-Man is standing in a set position and needs to get something. He throws a web at the thing he wants, and then drags it to him. Now imagine that a very tiny Spider-Man is standing at one end of a cell. He throws a web at the chromosome he wants, the web attaches, and he pulls the chromosome to

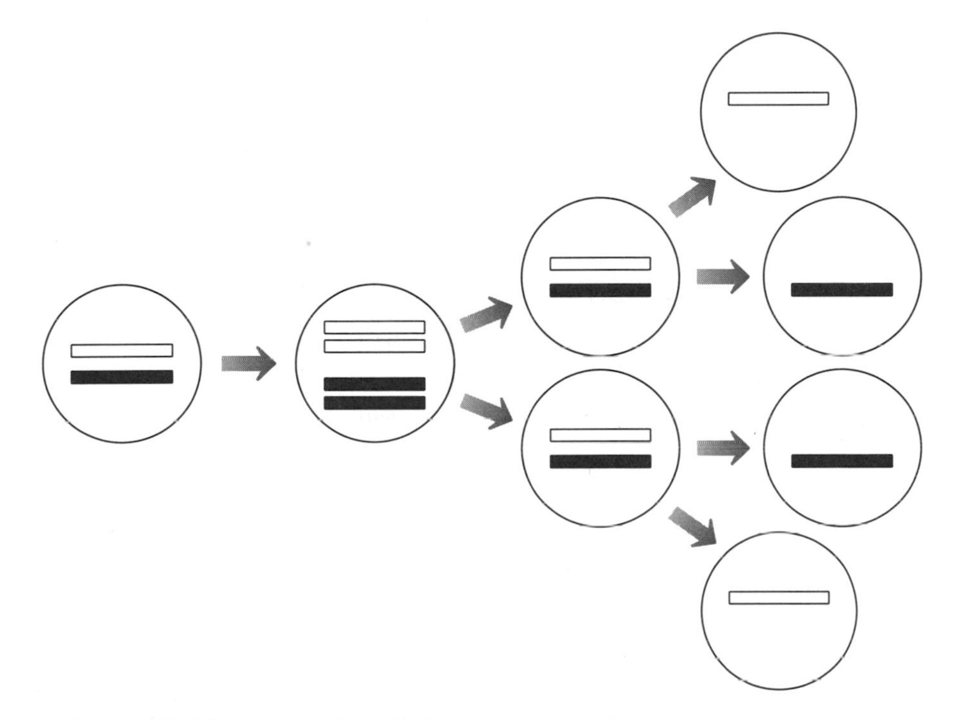

Figure 6.2 This shows the cell division process that generates gametes (eggs or sperm) each containing just one of every pair of chromosomes. The process initially looks like the standard cell division shown in Figure 6.1. However, this is followed by a second separation of chromosome pairs, to create gametes with only half the normal number of chromosomes. There is also an early event where genetic material is swapped over within chromosome pairs, to create greater genetic diversity in offspring, but this isn't shown in this figure.

his end of the cell. A tiny Spider-Man clone does the same thing at the opposite end of the cell for the other chromosome in the matching pair.

There is a complication for Spider-Man. Most of the surface of the chromosome is coated with web-repellent. There is only one part where his web will stick. This part is the centromere. In the cell the centromere attaches to a long string of proteins which pulls the chromosome away from the centre and to the periphery. This string of proteins is called the spindle apparatus.

Centromeres play a very important and consistent role in all species. They form the essential attachment point for the spindle apparatus. It's essential that this system works properly, or cell division goes wrong. Given that this is such a vital process, we would expect that the centromere DNA sequence would be highly conserved throughout the evolutionary tree. But weirdly, this isn't the case at all. Once we move beyond yeast* and microscopic worms,** the DNA sequence is highly variable when we look at different species.[2] In fact, the DNA sequence of a centromere may differ between two chromosomes in the same cell. This level of sequence diversity, in the face of functional consistency, is really quite counterintuitive. Happily, we are starting to understand how this vital region of junk DNA manages to pull off this strange evolutionary trick.

In human chromosomes, the centromeres are formed from repeats of a DNA sequence that is 171 base pairs in length.*** These 171 base pairs are repeated over and over, and may reach lengths of up to 5 million bases in total.[3] The critical feature of the centromere is that it acts as a location for the binding of the protein called CENP-A (*Cen*tromeric *P*rotein-*A*).[4] The CENP-A gene is highly conserved between species, in contrast to the centromere DNA.

Our Spider-Man analogy might be useful again here in terms of understanding the apparent evolutionary conundrum we laid out earlier. Spidey's web can bind to CENP-A protein. It doesn't matter if the CENP-A protein is bound to meat, bricks, potatoes or lightbulbs. So long as the CENP-A protein is bound to something, Spider-Man's web will stick to it, and pull the CENP-A and the something towards our superhero.

---

* Specifically, budding yeast such as *Saccharomyces cerevisiae*.

** *Caenorhabditis elegans*.

*** This unit of 171 base pairs is called an alpha (α) satellite repeat.

So, the DNA sequence at the centromere can vary enormously between species, ranging from meat to lightbulbs. What matters is that the CENP-A protein remains the same, so that the highly conserved spindle apparatus can stick to it and pull the chromosomes apart to opposite poles of the dividing cell.

CENP-A isn't the only protein that is found at the centromere; many others are also present. It's possible to knock out the expression of CENP-A in cells in the laboratory. When this happens, the other proteins that should bind to the centromere stop doing so.[5,6] However, when the experiment is performed the other way around – knocking out expression of one of the other proteins – CENP-A continues to bind at the centromere.[7] This demonstrates that CENP-A acts as a foundation stone.

When researchers over-expressed CENP-A in cells from fruit flies, they found that the chromosomes began to create centromeres in unusual positions.[8] But the situation in human cells seems to be more complicated, because over-expression of CENP-A doesn't result in new, abnormally located centromeres.[9] It seems that in humans, CENP-A is necessary for centromere formation, but it's not sufficient.

The CENP-A acts as the essential cornerstone for the recruitment of all the other proteins that are also required for the spindle apparatus to do its job. When a cell is actively dividing, over 40 different proteins build up from the CENP-A. They do so in a step-wise fashion, like adding on LEGO bricks in a particular order. Immediately after the duplicated chromosomes have been pulled to the opposite ends of the cell, this big complex falls apart again. This whole process can take less than an hour. We don't know what controls all of this, but some of it is down to a simple physical feature. Normally, the nucleus has a membrane around it, and large protein molecules find it really difficult to get through this. When the cell is ready to separate its replicated chromosomes, this barrier breaks down temporarily and the proteins can join on to the complex at the centromere.[10] It's like having a removal company outside your house. They are ready to shift your furniture but can't get on with the job unless you open the door and let them in.

Location, location, location

We are still left with a difficult conceptual problem. If the DNA sequence at the centromere isn't very conserved, and the critical factor is the placement of the CENP-A protein, how does the cell 'know' where the centromere should be on each chromosome? Why is it always near the middle of chromosome 1, but near the end of chromosome 14?

To understand this, we have to develop a more sophisticated image of the DNA in our cells. The DNA double helix is an iconic image, probably the defining image in biology. But it doesn't really represent what DNA is like. DNA is a very long spindly molecule. If you stretched out the DNA from one human cell it would reach for two metres, assuming you joined up the material from all the chromosomes. But this DNA has to fit into the nucleus of a cell, and the nucleus has a diameter of just one hundredth of a millimetre.

This is like trying to fit something that is the vertical height of Mount Everest into a capsule the size of a golf ball. If you are trying to fit a climbing rope the height of Mount Everest into a golf ball, that clearly won't work. On the other hand, if you replace the climbing rope with a filament thinner than a human hair, you'll probably be OK.

Although human DNA is long, it's very thin, so it is possible to fit it into the nucleus. But there is, as always, a complication. It's not enough just to jam the DNA into a small space. The easiest way to visualise why not is to think of strings of Christmas tree lights. If at the end of the festive season you take the lights off the tree and shove them into a box, they will take up a lot of space.

You will also almost certainly find that when you come to use them again the following year they are all tangled up. It will take you ages to unravel them and there is a fair chance you will break some of them. In their tangled state, you would also really struggle to get to just one particular bulb.

But, if you are a freakishly organised person, you will wrap each string of lights around a piece of cardboard before storing them away. And your organisational acumen will be rewarded next Christmas when you take the lights out of the surprisingly small box you were able to use for storage. Not only did you save on loft space, you also will find that it's very easy to unwind the lights, none of the strands get tangled around each other or snapped, and you can access your one favourite bulb very easily.

The same process happens in our cells. DNA is not stored as a random bundle of scrunched-up genetic material. Instead it is wrapped around certain proteins. This stops the DNA getting tangled and broken, allows it to be squeezed in an orderly fashion into a small space, and also keeps it structured so that the cell can access different regions as necessary, in order to switch individual genes on or off.

The DNA in our cells is wrapped around particular proteins, called histones. The basic structure is shown in Figure 6.3. Eight histone proteins – two each of four different types – form an octamer. DNA wraps around this octamer, like a skipping rope around eight tennis balls. There are huge numbers of these octamers all along our genome.

CENP-A is a close cousin of one of these histone proteins, sharing much of the same amino acid sequence, but with some important differences. At the centromere, both copies of one of the standard histone proteins are missing,* and CENP-A is present in the octamer instead, as shown in Figure 6.4.[11] There are thousands

* These are called histone H3.

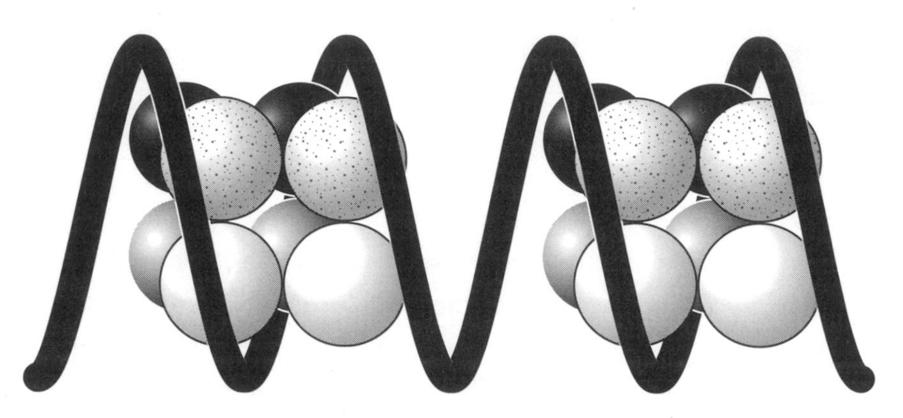

Figure 6.3 DNA, represented by the solid black band, is wrapped around packages of eight histone proteins (two each of four different types).

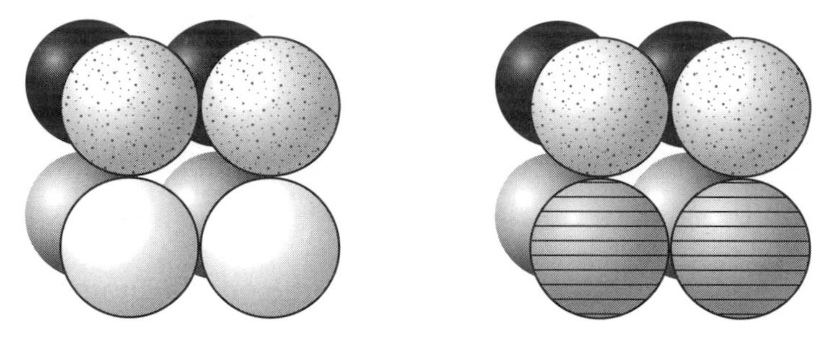

Figure 6.4 The octamer of histone proteins on the left represents the standard arrangement found throughout most of the genome. The octamer on the right represents the specialised octamers found at the centromeres. One of the standard pairs of histone proteins has been replaced by a pair of specialised centromere histone proteins, called CENP-A. These are represented by the striped globes.

of these octamers containing CENP-A at the centromere of each chromosome.

The CENP-A in these thousands of octamers at the centromeres gives the spindle apparatus something to hang on to, when it's trying to pull the chromosomes apart. One of the effects of inserting CENP-A into the octamers is that it makes the centromere regions more rigid.[12] If we think about trying to pull a blob of jelly, compared with a boiled sweet, it's obvious that the increased rigidity will be an advantage for the actions of the spindle apparatus.

But we still keep coming back to the same problem. Why is CENP-A inserted into the octamers at the centromere, but not at other regions? This isn't driven by the DNA sequence. Other regions of our genome also contain junk DNA with similar sequences to those found at the centromeres, but CENP-A doesn't accumulate at these.[13] CENP-A is only found at centromeres, but in some ways it's the presence of CENP-A that actually defines what a centromere is. How have human cells evolved in such a way that an inherently unstable situation has led to complete genetic stability in terms of cell division?

The answer lies in a self-seeding paradigm, whereby once CENP-A is deposited it continues to direct the maintenance of its own position, and to ensure that this is passed on to all daughter cells.[14] This is independent of DNA sequence. Instead it seems to depend on small chemical modifications to the histone octamers.

Histone proteins in the octamers can be modified in a huge number of different ways. Proteins are made up of combinations of 20 different amino acids, many of which can be modified. And there are lots of different modifications that can be made to a protein. This is just as true of histones as of any other proteins.

In human centromeres, the octamers that contain CENP-A don't have a complete monopoly. Instead, blocks of these octamers alternate with ones containing the standard histone protein, as shown in Figure 6.5. The standard octamers carry a very characteristic combination of chemical modifications. These in turn attract other proteins that bind to these modifications, part of whose function is to make sure these modifications are maintained.[15] This all acts to keep the octamers that contain CENP-A localised to the same region of the genome, and means that they only form at

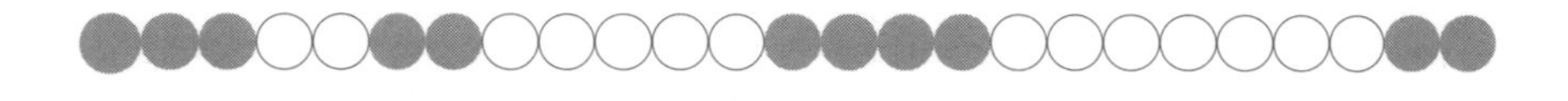

Standard octamer   Octamer containing CENP-A

Figure 6.5 The alternating pattern of standard and CENP-A histone octamers at the centromeres. For clarity, only small numbers of octamers are shown, whereas there are thousands present in the cell. Each circle represents an entire octamer.

one position on the chromosome. This is probably why the junk DNA sequence at centromeres is so variable between species, even though it provides the geographical scaffold for one of the most fundamental processes in any cell.

The chemical modifications at the centromere also have the effect of keeping that region of the genome silent. Although there are recent data suggesting that there may be low-level expression of RNA from some centromeric regions, it's very unclear if this has any functional significance. Essentially, the DNA at the centromeres has no real function except to be junk. It just acts as the regions where CENP-A and all its associated proteins can bind. That's the only thing the cell needs from it. It's better that it doesn't have any other purpose, because that might be disrupted when the octamers containing CENP-A bind. That's why this region of DNA has been able to change so much during evolution, because the sequence really doesn't matter.

Nothing comes from nothing

It might seem that there is still a missing stage in this. How does the CENP-A 'know' to bind to the right region of junk DNA in the first place? Because that tends to be how we all think, wanting to know what starts something off. But if we examine that assumption, we realise it leads us into a dead-end. Once again in this chapter we can invoke the lyricist Oscar Hammerstein, although this time in Austria rather than Siam/Thailand.

In *The Sound of Music*, Captain von Trapp and Maria sing that 'Nothing comes from nothing. Nothing ever could'.[16]

How right they were.

Naked human DNA is a completely non-functional molecule. It does nothing at all, and certainly can't direct the production of a new human being. It needs all the accessory information, such as the histones and their modifications, and it needs to be in a functioning cell. When the replicated chromosomes are separated and pulled to opposite ends of the cell, they each carry off some histone octamers in the correct positions, and with appropriate modifications. There are enough of these that they can act as the seed region to recreate the full picture of histones and modifications in the daughter cells. This is true not just of standard histone octamers, but also of the ones that contain CENP-A and thus show where the centromeres are formed. For these non-standard octamers, the regions of the CENP-A protein that contain different amino acids from the standard histones are important for attracting the appropriate proteins.[17]

This information – the chemical modifications – is even retained when eggs and sperm are produced.[18] The octamers that contain CENP-A stay in place when the egg and sperm fuse to form the one cell that will ultimately give rise to all the trillions of others in the human body. Our centromeres have been passed down through all of human evolution, and long before that in our distant ancestors, based on the position of the proteins, and not the DNA sequence to which they bind.

There are drugs that interfere with the way in which the spindle apparatus pulls the replicated chromosomes to opposite ends of the cells. The spindle apparatus is formed by the coming together of a large number of proteins, and these only combine at the time when a cell is ready to pull the chromosomes apart. A drug called paclitaxel works by making the spindle apparatus too stable, so that the complex of proteins can't disaggregate.[19]

We can visualise why this is a bad thing for a cell by comparing the scenario with one of those fire engines that carries an extending ladder. It's great that the ladder can be extended to rescue people from upper storeys of a burning building. But if the fire crew can't get the ladder folded back down again after the emergency and have to drive around with it fully extended, it won't be long before they have a pretty serious accident. The same happens in the cells treated with paclitaxel. Systems in the cell recognise that the spindle apparatus hasn't been deactivated properly, and this triggers destruction of the cell. In the UK, paclitaxel is licensed for use in a number of cancers including non-small cell lung cancer, breast cancer and ovarian cancer.[20]

Paclitaxel is probably effective because cancer cells divide rapidly. By using a drug that targets cell division, it's possible to kill the cancer cells at a higher rate than the normal body cells, which are not proliferating so quickly. But we also know that abnormal separation of chromosomes is itself a hallmark of many cancers.

The numbers matter

If the separation of chromosomes goes wrong, one daughter cell may inherit both the 'original' chromosome and its replicate. The other daughter cell won't inherit either. The first daughter cell will have one chromosome too many, the other daughter cell will have one too few. This situation, where the number of chromosomes is wrong, is known as aneuploidy. The word is derived from Greek. In this case, *an* means 'not', *eu* means 'good' and *ploos* means '-fold' (as in 'twofold', 'threefold', etc.). In other words, it represents an unbalanced genomic state.

Astonishingly, about 90 per cent of solid tumours contain cells that are aneuploid, i.e. contain the wrong number of chromosomes.[21] The pattern of aneuploidy can be really complicated, as there is probably a strong degree of randomness to how the chromosomes are mis-segregated if the process is going wrong. In a single cancer cell there may be four copies of one chromosome, two copies of another and one copy of a third, or some other combination. Because of this variability, it's very difficult to determine if the aneuploidy itself drives the cancer process, or if it's just an innocent marker of the cancer status of the cells. The likelihood, because of the essentially random patterns of abnormal chromosome numbers, is that there's probably a spectrum. Some cancer cells may develop combinations of chromosomes that drive cell proliferation faster. Other cells may have combinations with the opposite effect, and which may even trigger the cancer cell's suicide system. And in some cells the combination may be ultimately neutral.[22]

Remarkably, aneuploidy also seems to occur in certain normal cells. It's been reported that perhaps as many as 10 per cent of cells in the brains of mice and humans are aneuploid.[23] During development, the proportion is even higher, at around 30 per cent, but many of these are eliminated.[24] As far as we can tell, the remaining aneuploid cells in the brain are functionally active.[25] There is no clear understanding of why we have these brain cells with abnormal numbers of chromosomes, or the significance of similar findings of aneuploidy reported in the liver.[26]

In the situations outlined above, the aneuploidy has developed after the main bulk of the cells of the body have been produced. It occurred during cell divisions that were creating new body cells, albeit in some cases cancerous ones. The effects of these failures in chromosome segregation seem relatively mild, if any. That's probably because there are plenty of normal cells to compensate.

But the situation is very different if the aneuploidy occurs during the formation of the eggs or sperm (gametes). If a pair of chromosomes fails to separate properly, then one of the resulting gametes will have an extra copy of the chromosome, and the other will be lacking that chromosome. Let's say that happens in the formation of the egg, and chromosome 21 is abnormally segregated when the eggs are created. One of the eggs will have two copies of chromosome 21, the other will have none.

If the one that lacks a chromosome 21 is fertilised, the resulting embryo only has one copy of chromosome 21 and very quickly dies. But if the egg that contains two copies of chromosome 21 is fertilised, it will have three copies of this chromosome. And although such embryos are at higher than normal risk of spontaneous abortion, many do develop fully and the child is born.

Most of us have met or at least seen people with three copies of chromosome 21 (having three copies is known as a trisomy, so this condition is known as trisomy 21): this failure of chromosome segregation is the cause of Down's Syndrome.[27] It can also occur because of a sperm with two copies of the chromosome, or through failure of chromosome separation in the first few divisions after fertilisation, but the maternal route is the most common.

Down's Syndrome affects about one in 700 live births, and is a complex and variable disorder commonly associated with heart defects, a characteristic physical and facial appearance and a greater or lesser degree of learning disability. People with Down's Syndrome are much more likely to reach adulthood than in the past, thanks to better medical and surgical interventions, but are at high risk of a relatively early onset of Alzheimer's disease.[28]

The complex nature of the characteristics of Down's Syndrome demonstrates very clearly that it's really important that our cells contain the correct number of chromosomes. Patients with Down's Syndrome have three copies of chromosome 21 instead of two. But this 50 per cent increase in the chromosome number, and therefore of the genes on the chromosomes, has dramatic effects on the cell and on the individual. Our cells are simply unable to deal with this excess, showing that control of gene expression must normally be tightly regulated and is so finely balanced that we are only able to compensate for changes within relatively narrow parameters.

Two other trisomies have been found in humans, both associated with much more severe conditions than Down's Syndrome. Edward's Syndrome is caused by trisomy of chromosome 18, and affects one in 3,000 live births. Approximately three-quarters of foetuses with trisomy 18 die in utero. Of the babies who survive to term, about 90 per cent die in the first year of birth due to cardiovascular defects. The babies grow very slowly in the womb, their birth weight is low and they have a small head, jaw and mouth plus a range of other multisystem problems including severe learning disabilities.[29]

The rarest of all these conditions is Patau's Syndrome, trisomy 13, which affects one in 7,000 live births. The babies who survive to full term have severe developmental abnormalities and rarely survive their first year. A wide range of organ systems is involved, including the heart and kidneys. Severe malformations of the skull are common and the learning disability is extremely severe.[30]

It's notable that having an extra chromosome from conception onwards results in obvious developmental problems. In each of these trisomies, it is very clear that the baby has a major problem from the moment they are born. Indeed, with access to prenatal scanning, most of the affected foetuses are detected during pregnancy. This tells us that having the right dose of chromosomes is vitally important for the highly coordinated process of development.

It's tempting to wonder if there is something unusual about chromosomes 13, 18 and 21. Is there, perhaps, something different about their centromeres that makes them more susceptible to unequal segregation of the chromosomes during the formation of the egg and the sperm? Or could it be that trisomies of the other chromosomes do occur, but there's no clinical effect so we don't think to look for them?

This is falling into the surprisingly common trap of focusing on what we see, rather than what we don't see. The reason that we see babies born with trisomies of chromosomes 13, 18 and 21 is because these are relatively benign, unlikely though that sounds. These are three of the smallest chromosomes and they each contain relatively few genes. Generally, the larger the chromosome, the greater the number of genes it contains. So the reason we never see trisomy of chromosome 1, for example, is because of its size. Chromosome 1 is very large and contains a lot of genes. If an egg and sperm fuse and create a zygote with three copies of this chromosome, there will be overexpression of such a large number of genes that the cell function will be disrupted catastrophically, leading to extremely early destruction of the embryo. This probably occurs before the woman is even aware she is pregnant.

For women aged between 25 and 40, the success rates for in vitro fertilisation using donated eggs are not affected by age.[31] But the likelihood of a woman becoming pregnant naturally does decline after her mid-20s. The difference between these two situations suggests that the mother's age critically affects her eggs, rather than her uterus. We already know from Down's Syndrome that maternal age influences the success of chromosome segregation into the eggs. So it's not too big a leap to hypothesise that the decline in pregnancy rates after the mid-20s may be in part due to very early failures of embryo development, as a result of malfunctioning centromere activity and the creation of eggs with disastrous misallocation of large chromosomes.

7. Painting with Junk

In a twelve-month period from 2011 to 2012, 813,200 babies were born in the UK.[1] Using the rates quoted in the previous chapter, we can estimate that nearly 1,200 of these babies had Down's Syndrome, around 270 had Edward's Syndrome and just under 120 had Patau's Sydrome. That's a very small number of cots in a nursery of over three-quarters of a million babies. This is consistent with the concept that having too many copies of a chromosome is very damaging: in general we would not expect high survival rates when it occurs.

Which makes it all the more surprising to learn that about half of the babies born in that period – that's over 400,000 children – were born with one chromosome too many. Yes, one in two of us. Even more confusingly, the extra chromosome isn't some tiny little genetic remnant. It's a really big chromosome. How on earth can this be, when one extra copy of a very small chromosome can cause devastating conditions such as Edward's or Patau's Syndromes?

The culprit here is known as the X chromosome, and it's prevented from causing harm by a process that relies utterly on junk DNA. But before we move to exploring how this protection happens, we need to explore the nature of the X chromosome itself.

Most of the time the chromosomes in a cell are very long and stringy, and difficult to distinguish from each other. They appear like a great bundle of tangled wool when viewed under a normal light microscope. But when a cell is getting ready to divide, the chromosomes become very structured and compact, and are really discrete entities. If you know the right techniques, you can isolate all the compacted chromosomes from a nucleus, stain them with specific chemicals and examine the individual ones through a microscope. At this stage they look more like separate skeins of embroidery wool, with the centromere as the little tube of paper that holds the skeins in place.

By analysing photos of the whole complement of chromosomes in a human cell, scientists were able to identify each individual chromosome. They literally used to cut and paste the individual chromosome pictures to arrange them in order. This is how researchers discovered the causes of Down's, Edward's and Patau's Syndromes, by analysing the chromosomes in cells taken from affected children.

But before identifying the underlying problems in these serious conditions, the early researchers discovered the fundamental organisation of our genetic material. They showed that the normal number of chromosomes in a human cell is 46. The exceptions are the eggs and the sperm, which each have 23. Our chromosomes are arranged in pairs, inherited equally from our mother and father. In other words, one copy of chromosome 1 from mum and one from dad. The same for chromosome 2, and for the others.

This is true for chromosome 1 up to chromosome 22. These are known as the autosomes. If we only looked at the autosomes in a cell, we would not be able to tell if the cell was from a female or a male. But this information becomes immediately apparent if we look at the last remaining pair of chromosomes, known as the sex chromosomes. Females have two identical large sex chromosomes, known as X. Males have one X chromosome and a very small chromosome, called Y. These two situations are shown in Figure 7.1.

The Y chromosome may be small, but it has an amazing impact. It's the presence of the Y chromosome that determines the sex of the developing embryo. It only contains a small number of genes, but these are vitally important in governing gender.

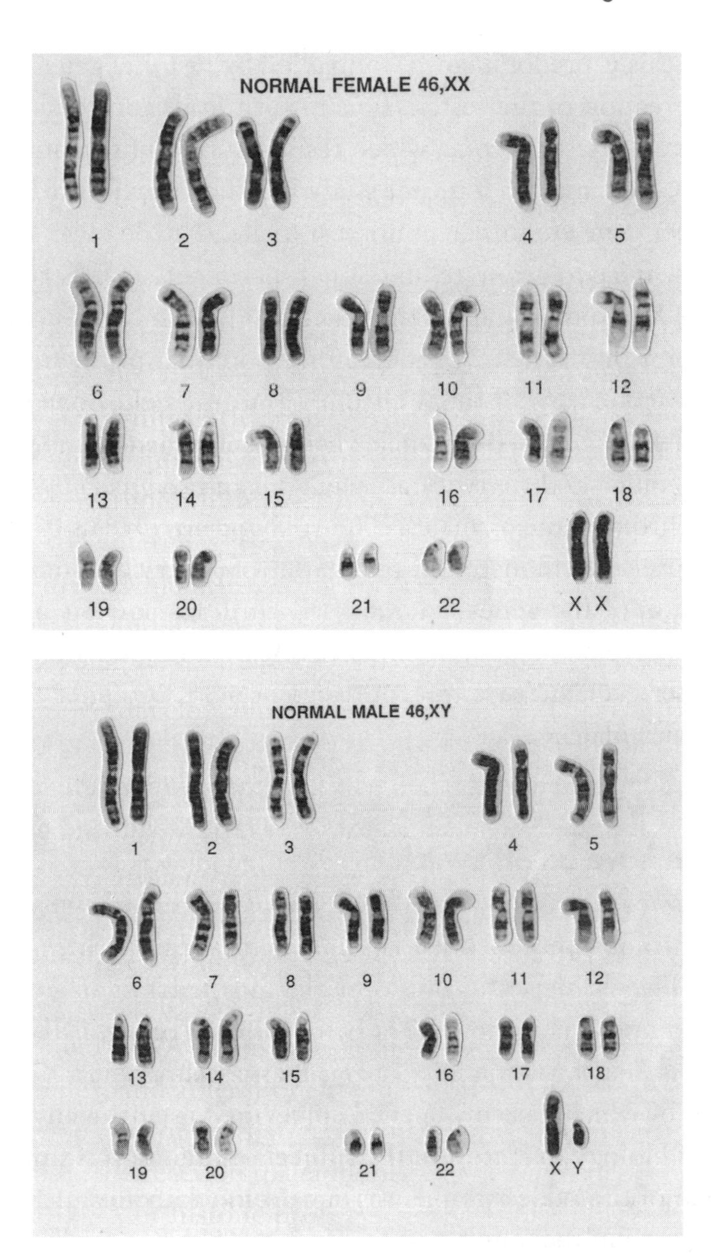

Figure 7.1 Standard female and male karyotypes, showing all the chromosomes present in a cell. The upper panel shows a female karyotype, the lower a male one. The only difference is in the last pair of chromosomes. Females have two large X chromosomes, males have one large X and a small Y. (Wessex Regional Genetics Centre, Wellcome Images)

In fact, this is predominantly controlled by just one gene*[2] which drives creation of the testes. This in turn leads to production of the hormone testosterone, which results in masculinisation of the embryo. Remarkably, a recent study has shown that just this and one other gene are sufficient not just to create male mice, but also for these mice to generate functional sperm and to father pups.[3]

The X chromosome, on the other hand, is very large, containing over 1,000 genes.[4] This creates a potential problem. Males only have one copy of the X chromosome and hence one copy of each of these genes. But females have double that number, so in theory could produce twice as much of the products encoded by the X chromosome as males. The trisomic conditions described in Chapter 6 demonstrated that even a 50 per cent increase in expression of the genes from a small chromosome has a hugely detrimental effect on development. How then can females tolerate a 100 per cent increase in expression of over 1,000 genes, compared with males?

Women have an off switch

The answer is that they don't. Females produce the same amount of X chromosome-encoded protein expression in their cells as males. They achieve this by a remarkably ingenious arrangement whereby one X chromosome is switched off in every cell. This is known as X-inactivation. Not only is it essential for human life, the process by which it occurs opened up new and totally unanticipated areas of biology that are still the subject of intense scrutiny.

One of the oddest things we have come to realise is that our cells can count the number of X chromosomes. Male cells contain an X and a Y chromosome and they never inactivate the single X. But sometimes males are born who have two X chromosomes and

---

* This gene is called *SRY*.

one Y. They are still males, because it's the Y chromosome that drives masculinisation. But their cells inactivate the extra X, just as female cells do.

A similar thing happens in females. Sometimes females are born who have three X chromosomes in each cell. When this happens, the cells shut down two X chromosomes instead of one. The flip side of this is when females are born who only have one X chromosome. In this case, the cell doesn't shut it off at all.

In addition to being able to count, our cells are also able to remember. When a female produces eggs, she usually only gives each egg one of each pair of chromosomes, including the X chromosome. A male produces sperm that contain either an X or a Y chromosome. When a sperm that contains an X chromosome fuses with an egg, the resulting single-cell zygote contains two X chromosomes and both are active. But very early in development, after just a few rounds of cell division, one X chromosome is inactivated in each cell of the embryo. Sometimes it's the X that came from father, sometimes the X that came from mother. Every daughter cell that subsequently develops switches off the same X chromosome as its parental cell. This means that of the 50 trillion or so cells in the adult female body, on average about half will express the X chromosome that was provided by the egg, and the other half will express the X chromosome that was provided by the sperm.

When an X chromosome is inactivated, it adopts a very unusual physical conformation. The DNA becomes incredibly compacted. Imagine you and a friend each take hold of opposite ends of a towel. You start turning your end of the towel clockwise, and your friend does the same at the other end. Pretty quickly, the towel will start twisting in the middle, and the two of you will be pulled closer together. Now imagine that the towel is about five metres in length, but you manage to keep twisting it until it's a dense clump of towel only a millimetre in linear length. By this stage, the towel is extraordinarily tightly wound up. Essentially, the X chromosome becomes as tightly compacted as that towel. One of the consequences is that it forms a dense structure that can be seen when looking at the nucleus of a female cell down a microscope, when all the other chromosomes are long and stringy and can't be visualised. The condensed X chromosome is called the Barr body.

In order to try to understand how X chromosome inactivation happens, scientists studied unusual cell lines and mouse strains. These focused on examples where parts of the X chromosome had been lost, or where bits of it had been transferred to other chromosomes. Some cells that had lost part of the X chromosome were still able to inactivate one of their X chromosomes, as shown by the presence of the Barr body. But cells that had lost a different part of the X weren't able to form Barr bodies, showing that they hadn't inactivated a chromosome.

Where parts of the X chromosome had been transferred to other chromosomes, sometimes these abnormal chromosomes were inactivated, and other times they weren't. It all depended on which bit of the X chromosome had been transferred.

These data enabled researchers to narrow down the region on the X chromosome that was key for inactivation. Rationally enough, they called this region the X inactivation centre. In 1991, a group reported that this region contained a gene that they called Xist.* Only the Xist gene on the inactive chromosome expressed Xist RNA.[5,6] This made perfect sense, because X inactivation is an asymmetric process. In a pair of equivalent X chromosomes, one is inactivated and one is not. So it seemed consistent that this process would be driven by a scenario where one chromosome expresses a gene and the other doesn't.

---

* The name Xist is derived from X-inactive (Xi)-specific transcript.

A very large bit of junk

It was obvious that the next question would be to ask how Xist works and the first thing that researchers did was to try to predict the sequence of the protein that it produced. This is usually relatively straightforward. Once they had found the sequence of the Xist RNA molecule, all that the scientists had to do was run this through a simple computer program that would predict the encoded amino acid sequence. Xist RNA is very long, about 17,000 bases. Each amino acid is encoded by a block of three bases, so a 17,000-base RNA could theoretically code for a protein of over 5,700 amino acids. But when the Xist RNA sequence was examined, the longest run of amino acids was just under 300. This was despite the fact that the Xist RNA was spliced, in the way we first saw in Chapter 2, so had lost all the intervening junk sequences.

The 'problem' was that the Xist RNA was liberally scattered with sequences that don't code for amino acids, but which act as stop signals when protein chains are being built up. We can envisage this as being a little like trying to build a tall tower out of LEGO. It is perfectly straightforward until someone hands you one of those roof bricks that doesn't have any of the attachment nodes on the top. Once you insert this brick, your tower can't get any bigger.

If Xist did encode a protein, it would seem very odd that a cell would go to the effort of creating an RNA that was 17,000 bases* in length just to produce a protein that could have been encoded by an RNA of about 5 per cent of that length. Researchers in the field realised relatively quickly that this wasn't what was happening. The reality was much stranger.

* Bases rather than base pairs, because RNA is single-stranded.

DNA is found in the nucleus. It's copied to form RNA, and messenger RNA is transported out of the nucleus to structures where it acts as a template for protein assembly. But analyses showed that Xist RNA never left the nucleus. It doesn't encode a protein, not even a short one.[7,8]

Xist was in fact one of the first examples of an RNA molecule that is functional in its own terms, not as a carrier of information about a protein. It's a great example of how junk DNA – DNA which doesn't lead to production of a protein – is anything but junk. It's extremely important in its own right, because without it X inactivation cannot happen.

An odd feature of Xist is not just that it doesn't leave the nucleus. It doesn't even leave the X chromosome that produces it. Instead, it essentially sticks to the inactive X and then spreads along the chromosome. As more and more Xist RNA is produced, it begins to spread out and cover the inactive X chromosome, in a process quaintly referred to as 'painting'. The fact that this rather descriptive term is used is a quite good indicator that it's something we don't particularly understand. No one really knows the physical basis of how the Xist RNA creeps along the chromosome, like the mile-a-minute vine covering a wall. Even after more than twenty years we are still pretty hazy on how this happens. We do know that it's not based on the sequence of the X chromosome. If the X inactivation centre is transferred on to an autosome in a cell, then the autosome can be inactivated as if it were an X.[9]

Although Xist is required to initiate the process of X inactivation, it has helpers that strengthen and maintain the process. As Xist paints the X chromosome, it acts as an attachment point for proteins in the nucleus. These bind to the inactivating X, and attract yet more proteins, which shut down expression even more tightly. The only gene that isn't coated with Xist RNA and these proteins is the Xist gene itself. It remains a little beacon of expression in the chromosomal darkness of the inactive X.[10]

Left to right, right to left

So we have here a situation where a piece of 'junk' DNA – one that doesn't code for protein – is absolutely essential for the function of half the human race. Scientists have recently discovered that this process of X inactivation requires at least one other piece of junk DNA. Confusingly, this is encoded in exactly the same place on the X chromosome as Xist. DNA, as we know, is composed of two strands (the iconic double helix). The machinery that copies DNA to form RNA always 'reads' DNA in one direction, which we could call the beginning and end of a specific sequence. But the two strands of DNA run in opposite directions to each other, a little like one of those funicular railways we find at older seaside and mountain resorts. This means that a particular region of DNA may carry two lots of information in one physical location, running in opposite directions to each other.

A simple example in English is the word DEER, formed by reading from left to right. We could also read the same letters from right to left and in this case we would get the word REED. Same letters, different word, different meaning.

The other key piece of junk DNA involved in X inactivation is called, rather fittingly, Tsix. This is of course Xist spelt backwards, and it is found in the same region as Xist but on the opposite strand. Tsix encodes an RNA of 40,000 bases in length, over twice the size of Xist. Like Xist, Tsix never leaves the nucleus.

Although Tsix and Xist are encoded on the same part of the X chromosome, they are not expressed together. If an X chromosome expresses Tsix, this prevents the same chromosome from expressing Xist. This means that Tsix must be expressed by the active X chromosome, unlike Xist, which is always expressed from the inactive one.

This mutually exclusive expression of Tsix and Xist is of critical importance at a point in early development. The X chromosome in the egg has lost any of the protein marks that show it was inactivated (if it was the inactive version) and the X chromosome in the sperm had never been inactivated anyway. Following fusion and six or seven rounds of cell division, there will be a hundred or so cells in the embryo. At this stage, each cell in the female embryo switches off one of its two X chromosomes randomly. This requires a fleeting but intense physical relationship between the pair of X chromosomes in a cell. For just a couple of hours the two X chromosomes are physically associated in a brief encounter that ends with one being inactivated. The association is only over a small region of the X chromosome – the X inactivation centre, which codes for both Xist and TsiX RNA.[11]

A fleeting moment lasts forever

This is the mother of all one-night stands. In those two hours, chromosomal decisions get made which are then maintained for the rest of life. Not just during foetal development, but right up until the woman dies, even if that is more than a hundred years later. And it affects not just the hundred or so cells, but the trillions that come after them, because the same X chromosome is inactivated in all daughter cells.

It's still not entirely clear what happens during the hours of X chromosome intimacy in early development. The current theory is that there is a reallocation of junk RNA between the two chromosomes, such that one ends up with all the Xist and becomes the inactive X. We don't know how, but it's possible that one chromosome expresses slightly more or less of Xist or another key factor. We do know that the process begins just as levels of Tsix start to drop. It may be that once its levels fall below a certain critical threshold, Xist can start getting expressed from one of the X chromosomes.

Gene expression tends to have what's known as a stochastic component, by which we simply mean there's a bit of random variability in the levels. If one of the chromosomes is expressing a slightly higher amount of one or more key factors, this may be sufficient to build a self-amplifying network of proteins and RNA molecules. Because the inequality in expression is essentially stochastic (due to random 'noise') the inactivation will also be essentially random across the hundred or so cells.

Here's a possible way of visualising this. Imagine you get home late one evening and you have a hankering for melted cheese on two slices of toast. Just as you start to make this delicious supper, you realise you don't have much cheese in the fridge. What do you do? Make two rounds where neither really contains enough cheese to be satisfying? Or concentrate all of it on one slice, so that you get the dairy hit you are craving? Most people probably choose the latter, and in a way this is what the pair of X chromosomes do during the phase when random inactivation is taking place in the embryo. Evolution has favoured a process whereby, rather than each have a sub-critical amount of a key factor, the factor migrates to the chromosome that has slightly more to begin with. The more you have, the more you get.

X inactivation is entirely dependent on 'junk' DNA, and really gives the lie to that terminology. The process is absolutely essential in female mammals for normal cell function and a healthy life. It also has consequences in various disease states. Full-blown Fragile X syndrome of mental retardation, which we encountered in Chapter 1, only affects boys. This is because the gene is carried on the X chromosome. Women have two X chromosomes. Even if one of their chromosomes carries the mutation, enough protein is produced from the other (normal) one to avoid the worst of the symptoms. But males only possess one X chromosome and one Y chromosome, which is very small and doesn't carry many genes apart from the sex determining ones. Consequently, there is no compensatory normal Fragile X gene in males who carry a mutation on their X chromosome. If their sole X chromosome carries the Fragile X expansion, they can't produce the protein and so they develop symptoms.

This is also true of a whole range of genetic disorders where the mutated gene is carried on the X chromosome. Boys are more likely to have symptoms of an X-linked genetic disorder than girls, because the boys can't compensate for a faulty gene on their single X chromosome. Relevant medical conditions range from relatively mild issues such as red–green colour blindness to much more severe diseases. These include haemophilia B, the blood clotting disorder. Queen Victoria was a carrier of this condition and one of her sons (Leopold) was a sufferer and died at the age of 31 from a brain haemorrhage. Because at least two of Victoria's daughters were also carriers, and the royal families of Europe tended to inter-marry, this mutation was passed on to various other dynasties, most famously the Romanov line in Russia.[12]

Although women carrying the mutation that causes haemophilia only produce 50 per cent of the normal amounts of the clotting factor, this is enough to protect them from symptoms. This is partly because this clotting factor is released from cells and circulates in the bloodstream, where it reaches high enough levels for protection against bleeds, no matter where they happen.

There are, however, circumstances wherein the presence of two X chromosomes in a woman doesn't guarantee protection from an X-linked disorder. Rett Syndrome is a devastating neurological disease which presents in some ways as a really extreme form of autism. Baby girls appear to be perfectly healthy when born and they reach all the normal developmental milestones for the first six to eighteen months of life. But after that, they begin to regress. They lose any spoken language skills they have developed. They also develop repetitive hand actions, and lose purposeful ones such as pointing. The girls suffer serious learning disability for the rest of their lives.[13]

Rett Sydrome is caused by mutations in a protein-coding gene on the X chromosome.*[,14] Affected females have one normal copy of this gene, and one version which is mutated and can't produce functional protein. Assuming random X inactivation, we expect that on average half of the cells in the brain will express normal amounts of the protein, and there will be no expression from the other ones. It is obvious from the clinical presentation that there are severe problems if half the brain cells can't express this protein.

Rett Syndome pretty much only affects girls. This is unusual for an X-linked disorder, where girls are usually carriers and boys are affected. This might make us wonder how boys are protected from the effects of a Rett mutation. But the reality is that they are not. The reason we almost never find boys who are affected by Rett Syndrome is because affected male embryos don't develop properly and the foetuses don't survive to term.

Never underestimate luck, good or bad

Scientists are trained to think about many things during our education and careers. But something we are rarely asked to ponder is the role played by luck. Even when we do, we usually dress it up with terms like 'random fluctuations' or 'stochastic variation'. And that's a shame, because sometimes 'luck' is probably a better description.

Duchenne muscular dystrophy is a severe muscle wasting disease, which we first met in Chapter 3. Boys with this disorder are fine initially but during childhood their muscles begin to degenerate, in a characteristic pattern. For example, in the legs the thigh muscles begin to waste first. The boys develop very large calves as their bodies try to compensate, but after a while these muscles also

* The gene is called MeCP2 and its role is to bind to epigenetically modified (methylated) DNA, where it interacts with other proteins and represses gene expression at the sites where it binds.

wither. The children are usually wheelchair users by their teens and the average life expectancy is only 27 years of age. The early mortality is caused to a large extent by the eventual destruction of the muscles involved in breathing.[15]

Duchenne muscular dystrophy is caused by a mutation in a gene on the X chromosome that encodes a large protein called dystrophin.[16] This protein seems to act as a sort of shock absorber in muscle cells. Because of the mutation, males can't produce functional protein and this ultimately leads to destruction of the muscle. Carrier females will usually produce 50 per cent of the normal amounts of functional dystrophin protein. This is generally sufficient, because of an odd anatomical feature. As we develop, individual muscle cells fuse to create a large super-cell with lots of individual nuclei in it. This means each super-cell has access to multiple copies of the necessary genes, in all the different nuclei. So the muscles of carrier females overall contain enough dystrophin protein for normal activity, instead of one cell with enough, and one cell with none.

There was an unusual case of a woman with all the classic symptoms of Duchenne muscular dystrophy. This is very rare but there are ways we could predict this would happen. One possibility would be if her mother was a carrier and her dad was a Duchenne sufferer who survived long enough to father a child. If that was the case she would definitely have inherited a mutated gene from her father (because he would only possess one – affected – X chromosome). There would be a one in two chance that any egg produced by her carrier mother also contained a mutated dystrophin gene. If that scenario had occurred, neither of her X chromosomes would have a normal copy of the gene, and she wouldn't be able to produce the necessary protein.

But the doctors treating this patient had taken a family history and they knew that her father didn't have Duchenne muscular dystrophy, so another explanation was necessary. Sometimes mutations arise quite spontaneously when eggs or sperm are produced. The gene that codes for dystrophin is very large, so just by chance it is at relatively high risk of mutation compared with most other genes in the genome. That's because mutation is essentially a numbers game. The bigger the gene, the more likely it is that it may mutate. So, one mechanism by which a female could inherit Duchenne muscular dystrophy is if she inherits a mutated chromosome from her carrier mother, and a new mutation in the sperm that fertilised the egg.

This would normally seem like quite a good bet for explaining why this female patient had developed this disorder. There was only one problem. The patient had a sister. A twin sister. An identical twin sister, derived from the very same egg and sperm. And her twin sister was absolutely healthy. No symptoms of Duchenne muscular dystrophy at all. How on earth could two women who were genetically absolutely identical differ so much with respect to a genetically inherited disorder?

Think back to those hundred or so cells that undergo X inactivation during early embryonic development. Just by chance, about 50 per cent of them will switch off one X chromosome, and the remainder will switch off the other one. The same pattern of X inactivation is passed on to all the daughter cells throughout life.

The sister with Duchenne muscular dystrophy was simply incredibly unlucky during this stage. Just by sheer chance, all the cells that would ultimately give rise to muscle switched off the normal copy of the X chromosome. This was the one inherited from her father. This meant that the only X chromosome switched on in her muscle cells was the faulty one from her carrier mother. So none of the affected twin's muscle cells were able to express dystrophin and she developed the symptoms normally only seen in males.

When her genetically identical twin was developing, however, some of the cells that would give rise to muscle switched off the normal X chromosome and some switched off the mutated one. This meant that her muscles expressed enough dystrophin to keep them healthy, and she was an asymptomatic carrier, just like her mother.[17]

It is quite extraordinary to think that this was all caused by a simple fluctuation in the distribution of Xist, a long bit of RNA derived from junk DNA. The fluctuation lasted no more than a couple of hours, and occurred over a distance considerably less than one-millionth of the diameter of a human hair. Yet it was the difference between winning and losing in the health lottery.

Luck can be patchy

It is perhaps even stranger to think that some of the cat lovers among us look at, and stroke, the consequences of X inactivation every day. Tortoiseshell or calico cats (depending on which side of the Atlantic you're reading this book) are the ones with the distinctive patterns of orange and black. These coat colours occur in patches. The gene that controls the coat colour comes in two forms. An individual X chromosome carries either the orange version or the black version.

If the X chromosome carrying the black version is inactivated, the orange version on the other chromosome will be expressed and vice versa. When the cat embryo is at the size of a hundred cells or so, one or other X chromosome will be inactivated in each cell. And just as in all the other examples, all the daughter cells will switch off the same X chromosome. Eventually, some of these daughter cells will give rise to the cells that create pigment in the fur. As more and more of these cells divide and develop, they stay close to each other. This means that daughter cells tend to be clustered in patches. Because of the pattern of X inactivation in the daughter cells, this will lead to patches of orange fur and patches of black fur. This process is shown in Figure 7.2.

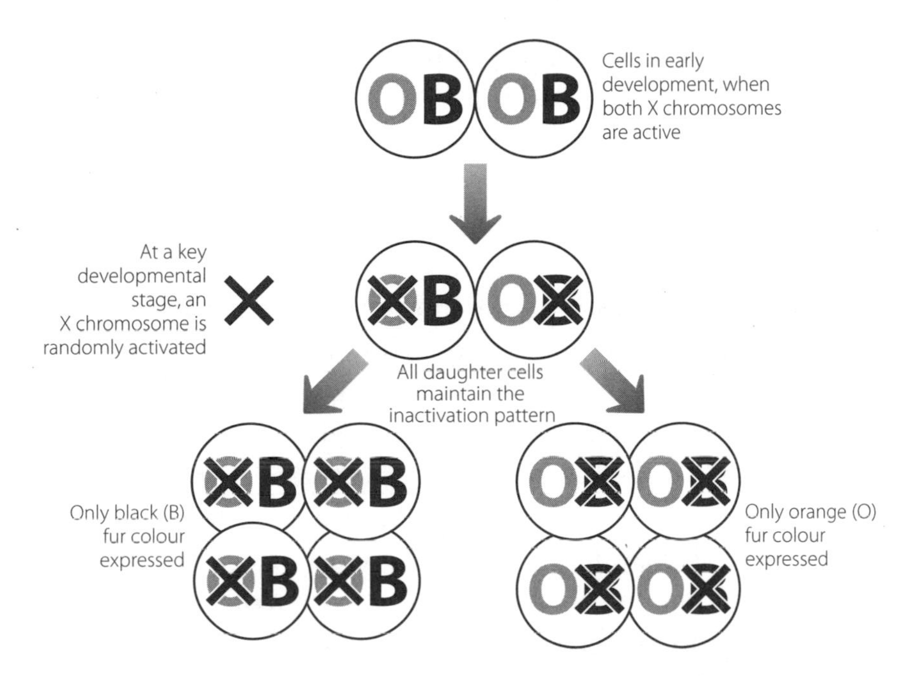

Figure 7.2 Schematic showing how patches of orange or black fur develop in female tortoiseshell cats depending on random X chromosome inactivation. The genes for fur colour lie on the X chromosome. If the black version is on the chromosome that is inactivated in a cell during early development, all descendants of that cell will only express the orange gene. The situation is reversed if the X chromosome carrying the orange gene is inactivated.

In 2002 scientists demonstrated beautifully just how random the process of X inactivation really is, by cloning a calico cat. They took cells from an adult female cat, and carried out the standard (but still fiendishly tricky) process of cloning. To do this, they removed the nucleus from the adult cat cell and put it into a cat egg whose own chromosomes they'd removed. This egg was implanted into a surrogate cat mother, and a lively and beautiful female kitten was born. And she didn't look anything like the genetically identical cat of which she was a clone.[18]

When this procedure is used to clone animals, the egg treats the new nucleus as if it was the real product of an egg fusing with a sperm. It strips off as much information as possible from the DNA, taking it back to its basic genetic sequence. This doesn't happen as effectively as in a real egg and sperm, which is one of the reasons why the success rate of this type of cloning is still very low. But sometimes it does work, as was the case here, and a cloned animal is born.

When the nucleus from the mother cat was put inside a cat egg, the egg caused changes to the chromosomes. One of these changes was the removal of the inactivating proteins on one of the X chromosomes, and the switching off of Xist expression. So for a short period in early development, both copies of the X chromosome were active. As the embryo developed, it went through the normal process at around the 100-cell stage of randomly inactivating an X chromosome in each cell. The pattern of X inactivation was passed on to daughter cells in the standard way, and the kitten thereby developed a different distribution of orange and black fur from its clonal 'parent'.

The moral of this story? If you have a calico cat you think is exceptionally beautiful, take lots of videos, lots of photos and if you want to be very weird about it, call in a taxidermist when she dies. But if you are ever approached by a door-to-door travelling cloner, just send them on their way.

8. Playing the Long Game

For quite a few years, Xist was considered an anomaly, a strange molecular outlier with an extraordinarily unusual impact on gene expression. Even when Tsix was identified, it was possible to think that junk RNAs were restricted to the vital but unique process of X inactivation. It is only in recent years that we have begun to recognise that the human genome expresses thousands of this type of molecule, and that they are surprisingly important in normal cellular function.

We now categorise Xist and Tsix as members of a large class known as the long non-coding RNAs. The term is a somewhat misleading one, because of course what it means is non-coding with respect to proteins. As we shall see, the long non-coding RNAs do code for functional molecules. The functional molecules are the long non-coding RNAs themselves.

Long non-coding RNAs are defined rather arbitrarily as molecules which are greater than 200 bases in length, and which don't code for proteins, making them different from messenger RNA. 200 bases is the lower size limit, but the biggest long non-coding RNAs can be 100,000 bases in length. There are lots of them, although no agreement yet on the precise number. Estimates range from 10,000 to 32,000 in the human genome.[1,2,3,4] But although there are a lot of long non-coding RNAs, they don't tend to be expressed to as high a level as the classical messenger RNAs which code for proteins. Normally, the expression level of a long non-coding RNA is less than 10 per cent of the level of an average messenger RNA.[5]

This relatively low abundance of any one long non-coding RNA is one of the reasons why we have tended to disregard this type of molecule until fairly recently. Essentially, when the expression of RNA molecules from cells was analysed, the long non-coding RNAs simply could not be detected very reliably because the technology wasn't sensitive enough. However, now that we know about their existence, we might think we should be able to analyse the genome of any organism, including humans, and predict their existence from the DNA sequence. We are, after all, pretty good at doing that for protein-coding genes.

But there are a number of aspects that make this difficult. We can identify putative protein-coding genes because of a number of features. They have certain sequences near the beginning and end of the genes that help us to find them. They also encode predicted runs of amino acids, which again give us confidence that a protein-coding gene may be present. Finally, most protein-coding genes are pretty similar if you look at a specific gene in different species. This means that if we identify a classical gene in an animal such as a pufferfish, it's easy to use that sequence as a basis for analysing the human genome to see if we can predict the presence of a similar gene in ourselves.

However, long non-coding RNAs don't have such strong sequence indicators as protein-coding genes, and they are also poorly conserved across species. Consequently, knowing the sequence of a long non-coding RNA in another species may not help us to identify a functionally related sequence in the human genome. Less than 6 per cent of a specific class of long non-coding RNAs in zebrafish, a common model system, have clearly equivalent sequences in mice and humans.[6] Only about 12 per cent of the same class of long non-coding RNAs that are found in humans and mice can be detected elsewhere in the animal kingdom.[7,8]

The relatively poor conservation of long non-coding RNAs was confirmed in a recent study comparing expressed long non-coding RNAs from various tissues of different tetrapod species. Tetrapod refers to all land-living vertebrates along with those that have 'returned to the sea' such as whales and dolphins. This paper reported that there were 11,000 long non-coding RNAs that were only found in primates. Only 2,500 were conserved across tetrapods, of which a mere 400 were classified as ancient, by which the authors meant that they had originated over 300 million years ago, around the time when amphibians and other tetrapods diverged. The authors suspected that the ancient long non-coding RNAs are the ones that are most actively regulated in all organisms, and are probably mostly involved in early development.[9] Most vertebrates look very similar during the earliest stages of embryogenesis, so it may make sense that we and all our distant cousins are using similar pathways to get started.

The generally poor conservation across species has led some authors to speculate that the long non-coding RNAs are not very important. The rationale behind this is that if they were significant they would be more constrained to remain similar during evolution and the development of species; whereas instead, the sequences coding for these 'junk' RNAs are evolving much more rapidly than the ones that encode proteins.

Although this is a fair point, it's perhaps an over-simplification. Long non-coding RNA molecules may be long in terms of the number of bases they contain, but that doesn't necessarily mean they are elongated stringy molecules in the cell. This is because long RNA molecules can fold onto themselves, forming three-dimensional structures. The bases in RNA pair up, following similar rules to the way in which the two strands of DNA are bonded together. RNA is a single-stranded molecule, so its bases pair up over relatively short distances, bending the molecule into complex stable shapes. These 3D structures may be very important in the function of the long non-coding RNA, and it's possible that the 3D structure is conserved across species, even if the base sequence is not.[10] This is shown in Figure 8.1. Unfortunately, predicting similar structures is difficult to do using sequence data, limiting the usefulness of this technique in helping us to find functionally conserved long non-coding RNAs.

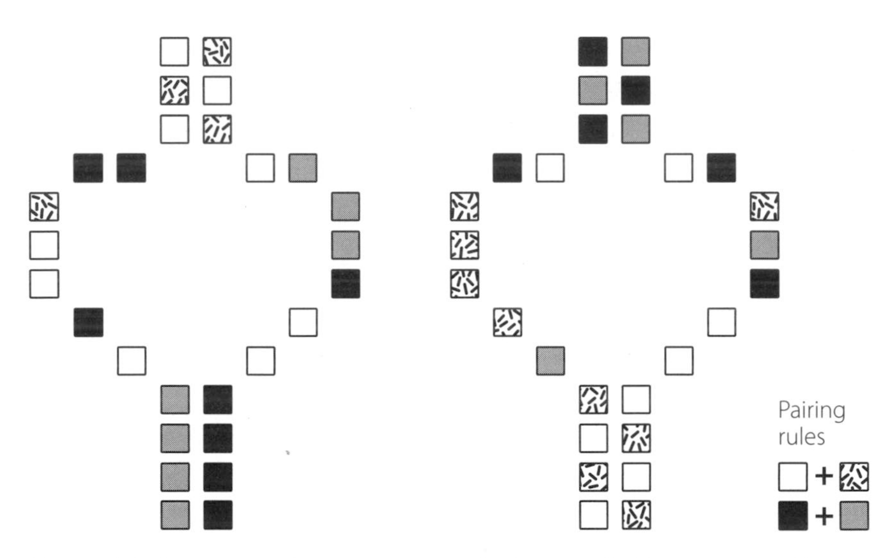

Figure 8.1 Representation of how two single-stranded long non-coding RNA molecules with different base sequences can form the same shape as each other. The shapes are determined by pairing of the A and U or C and G bases, which are represented by the differently shaded/patterned boxes. The representation is an over-simplification. In reality, the long non-coding RNAs may have multiple regions that can form complex structures. They will also be three-dimensional, rather than the flat shape shown here.

Logs or chips?

Because of the complications that arise if we try to identify long non-coding RNAs from the human genome sequence, most researchers lean towards the more pragmatic approach of identifying long non-coding RNAs by detecting the molecules themselves in cells. But there is a considerable degree of conflict in the scientific community about how to interpret the results. Hardcore junk aficionados might claim that if a sequence is expressed as a long non-coding RNA molecule then that molecule is being expressed for a reason. Other scientists are much more sceptical, positing that the expression of the long non-coding RNAs is essentially what we call a bystander event. This means that the long non-coding RNAs are expressed, but just as a by-product of switching on a 'proper' gene.

To understand what's meant by a bystander event, let's imagine we are cutting up tree branches with a chainsaw. The major aim of our activity is to create logs that we can use to build a cabin or to provide fuel for a stove. We aren't trying to create woodchips or sawdust, but this happens anyway as a result of the chainsaw function. It's not worth our while trying to avoid creating the woodchips. They don't really interfere with our main aim, and if we do find a way to avoid generating them, it might be at the expense of efficient production of the logs. Just occasionally, we may even find that we have a use for the woodchip by-product, using it to mulch a flowerpot, or provide bedding for our pet snake.

In a similar model, the junk sceptics postulate that expression of long non-coding RNA simply reflects a loosening of repression when genes in a particular region are expressed. In this model, the production of long non-coding RNAs is simply an inevitable consequence of an important process, but essentially harmless and insignificant. The believers counter that that fails to address certain aspects of long non-coding RNA expression. For example, different types of long non-coding RNAs are expressed if we examine samples from different brain regions.[11] Enthusiasts for long non-coding RNAs claim this supports their model for the importance of these molecules, because why else would different brain regions switch on different long non-coding RNAs? The sceptics claim that the different long non-coding RNAs are detected simply because various brain regions switch on different classical protein-coding genes. In our chainsaw analogy, this is equivalent to getting different woodchips depending on whether we are sawing up oak branches or pine.

It's early days but current data suggest that the extremists on both sides should probably relax a little because the reality is likely to lie somewhere between their two positions. The only way we can really test the hypothesis that long non-coding RNAs have functions in the cell is to test each one, in the correct cell type. Although perfectly sensible as an approach, this isn't as straightforward as it sounds. Partly this is down to sheer numbers. If we detect hundreds or thousands of different long non-coding RNAs in a cell or tissue, we have to make a decision about which one we want to test. But to do that, we already need to have developed a hypothesis about what that specific long non-coding RNA might do in the cell. Without that hypothesis, we won't know what effects we should be looking for if we interfere with the expression or function of that molecule.

Another complication is that many of the long non-coding RNAs are found in the same region as classical protein-coding genes. Sometimes they may be in exactly the same position, but encoded on the opposite strand, just as we saw for Xist and Tsix in Chapter 7. Others may be found within the stretches of junk that lie between two amino acid-coding regions in a single gene, which we first encountered in Friedreich's ataxia in Chapter 2 (*see page 18*). There are lots of ways in which the long non-coding RNAs may be co-located in the same region as protein-coding genes and this creates substantial experimental difficulties if trying to investigate function.

Usually the functions of genes are tested by mutating them. There are all sorts of mutations that can be introduced but the most commonly used will either switch the gene off or will lead to it being expressed at a higher level than normal. But because so many of the long non-coding RNAs overlap with protein-coding genes, it's hard to mutate one without mutating the other at the same time. We then face the problem of knowing whether the effects we see are due to the change in the long non-coding RNA or in the protein-coding gene.

A frivolous analogous example may help to visualise this problem. A PhD student was investigating how frogs hear. He had developed an experimental system where he surgically removed certain parts of a frog and then monitored if it could hear a loud noise, in this case a gunshot. One day he rushed in to his supervisor's office, yelling that he had worked out how frogs hear. 'They hear with their legs!' he told his bemused supervisor. When she asked how he could be so sure he said, 'It's simple. Normally if I fire the gun, the frog hears it and jumps in fright. But when I remove the frog's legs it doesn't jump anymore when I fire the gun, so it must hear through its legs.'*

Theoretically, of course, it's also possible that some of the unexpected effects sometimes encountered when we mutate protein-coding genes have been due to unrecognised changes in co-located long non-coding RNAs which we hadn't even realised were present at the time the experiment was carried out.

Because of this potential collateral damage to protein-coding genes, many researchers are focusing their efforts on a subset of long non-coding RNAs which don't overlap these regions. There's plenty of choice, as there are at least 3,500 long non-coding RNAs in this category. There is a tendency in the literature to refer to these more distant long non-coding RNAs as a special class, and they have been given a separate name.**[12]

---

* This is a famous thought experiment. No actual frogs were harmed in the creation of this anecdote.

** They are known as linc RNAs, which stands for long intergenic non-coding RNAs.

But it's worth remembering that if we do this, we are classifying these molecules by what they are not, i.e. they aren't co-located with protein-coding genes. This could mean that we lump together large numbers of long non-coding RNAs in one class when really they may turn out to be functionally quite distinct from each other.

The rush to create categories and nomenclature has been, and continues to be, a real problem in the whole field of genome analysis because it tends to lock us in to definitions before we really have enough biological understanding to create relevant categories. Imagine if you had never seen a movie, and then you were treated to a week of films. Let's imagine you see *Top Hat*; *Singin' in the Rain*; *The Good, the Bad and the Ugly*; *High Noon*; *The Sound of Music*; *The Magnificent Seven*; *Cabaret*; *True Grit*; *Unforgiven* and *West Side Story*. If asked to categorise movies, you would say they come in two flavours: musicals and westerns. That's fine, but what happens in the following week if you are shown *Bridget Jones's Diary* and *Gravity*? Or *Paint Your Wagon*, *Seven Brides for Seven Brothers* and *Calamity Jane*, all of which are song-and-dance films involving cowboys? You'll be stuck trying to shoehorn movies into genre definitions you developed before you understood the cinematic landscape. For a similar reason, we'll try to avoid too many definitions of individual classes of long non-coding RNAs and just focus on what we really know experimentally.

The importance of a good start in life

Appropriate control of gene expression is required throughout life, but it's critically important in very early development, because even the slightest shift in events during the first few cell divisions can have dramatic effects. This is particularly true in the zygote, the single cell formed from the fusion of an egg and a sperm.

The zygote, and the first few cells generated by division from this progenitor, are known as totipotent. They are able to create all the cells of the embryo and placenta. Researchers would love to work with these cells, but they are tiny in number. Instead, most research is carried out in embryonic stem cells, also known as ES cells. These were originally derived from embryos, many years ago, but we don't need to access embryos any more to get them, as they can be grown in cell culture. ES cells are from a slightly later stage in development and aren't quite as unconstrained as the zygote. They are known as pluripotent, as they have the potential to form any cell type in the body, but not placental cells.

In the correct, carefully controlled culture conditions, ES cells divide to generate yet more pluripotent stem cells. But relatively minor changes to the culture conditions lead to a loss of pluripotency. The ES cells begin to differentiate into more specialised cell types. One of the most dramatic changes is when ES cells differentiate into heart cells, which beat spontaneously and in synchrony in a Petri dish. But essentially the ES cells can move down many different development routes, depending on the ways that they are treated.

Researchers manipulated ES cells in culture by knocking down the expression of nearly 150 of the long non-coding RNAs that are located far from any known protein-coding genes. They knocked down the expression of just one long non-coding RNA in each experiment. They found that in dozens of cases, knockdown of just one long non-coding RNA was enough to change the ES cells from being pluripotent to starting to differentiate into other cells. The authors also analysed which genes were expressed before and after they knocked down the long non-coding RNAs. They found that over 90 per cent of the long non-coding RNAs controlled expression of protein-coding genes either directly or indirectly. In many cases, the expression of hundreds of protein-coding genes was affected. These were nearly always genes that were far away on the genome, not the ones that were closest to the long non-coding RNAs that they had knocked down.

The scientists also performed the reciprocal experiment. They treated ES cells with a chemical that is known to cause them to differentiate and then analysed the expression of the specific long non-coding RNA class in which they were interested. They found that expression of about 75 per cent of the long non-coding RNAs dropped as the cells moved from being pluripotent to being committed to a development pathway. The two sets of data are consistent with the idea that the levels of expression of certain long non-coding RNAs act as gatekeepers to maintain ES cells in a pluripotent state.[13] This created confidence that these non-protein-coding RNAs do have a function in the cell, at least during early development.

Some long non-coding RNAs may also affect later developmental stages. We met the HOX genes in Chapter 4. These are the genes that are important for correct patterning of body parts. They're the ones where mutations in fruit flies can lead to bizarre effects such as legs on the head. HOX genes are found in clusters in the genome, and these regions are extraordinarily rich in long non-coding RNAs. This is in contrast to their lack of ancient viral repeats. Scientists were keen to investigate if the long non-coding RNAs influenced the activity of the HOX genes in the same place in the genome. To test this, researchers used a technique to decrease the expression of a specific long non-coding RNA from the HOX gene region in chick embryos. When they did this, limb development went wrong. The bones towards the ends of the limbs were abnormally short.[14] Similarly, knocking out expression of another long non-coding RNA from this genome region in mice resulted in animals with malformations of the bones of the spine and wrists.[15] Both sets of data are consistent with the long non-coding RNAs being important regulators of HOX gene expression, and consequently of limb development.

Long RNAs and cancer

Cancer can in some ways be thought of as the flip side of development. One of the problems in cancer is that mature cells may change and revert to having some of the characteristics of less specialised cells, with a higher capacity to divide uncontrollably. Given that long non-coding RNAs are important in pluripotency and in development, it's perhaps not surprising that some have now been implicated in cancer.

One large study analysed the expression of long non-coding RNAs in over 1,300 individual tumours from four different cancer types (prostate, ovarian, a type of brain tumour called glioblastoma and a specific form of lung cancer). There were about 100 long non-coding RNAs where high levels of expression were most commonly found in patients who died quickly from the disease. Nine of these long non-coding RNAs showed this association no matter the class of cancer that was assessed, which suggests they may be useful as more general markers for predicting survival chances in a patient.[16]

For three of the cancer types (prostate cancer was the exception), the same study reported that they could detect long non-coding RNAs that differentiated one sub-class of tumour from another. Although we refer to ovarian cancer, for example, there are different types of ovarian cancer depending on the cell types involved, and this affects the natural history of the tumour in a patient. This in turn can have implications for the disease prognosis and the treatment that a patient should receive. Analysing the expression of specific long non-coding RNAs in a tumour sample may help clinicians in the future to select the most appropriate therapies for an individual patient.

The number of studies that report associations between long non-coding RNA expression and cancer are growing all the time. Intriguing data are also emerging from genetic studies of cancers. Some cancers are caused by a single really strong mutation which is passed on within a family. Probably the best-known example is the mutated BRCA1 gene which puts women at very high risk of aggressive breast cancer. It was knowing that she had a mutation in this gene that led the actress Angelina Jolie to elect for a double mastectomy in 2013. Such very strong single gene mutations are pretty rare in cancer. But studies have shown that quite a number of cancers do have a genetic component. The problem has been that when scientists mapped where the genetic variations were that were associated with cancer risk, they were frequently in regions of the genome where there were no protein-coding genes. Of just over 300 genetic variations linked to cancer, only 3.3 per cent changed amino acids in a protein, and over 40 per cent were located in regions between classical protein-coding genes. In these situations the variations may be affecting not protein-coding genes but long non-coding RNAs. Recent studies have confirmed this is the case for some of these variations in at least two cancer types (papillary thyroid cancer and prostate cancer).[17]

Encouragingly, we are also beginning to gather functional data that shows in some cases that these relationships are more than just associations, that the long non-coding RNAs are themselves causing alterations in the behaviour of the cancer cells.

There is a long non-coding RNA whose expression is increased in prostate cancer. This over-expression causes decreased expression of key proteins that normally hold cells back from proliferating too fast.[18,19] Over-expression of this long non-coding RNA is therefore essentially like releasing the handbrake on a car parked facing down a hill. The long non-coding RNA that causes skeletal deformations when it is knocked out in developing mice is over-expressed in a variety of cancers including liver,[20] colorectal,[21] pancreatic[22] and breast[23] and its over-expression is associated with poor prognosis for the patients. Studies using cancer cells in culture in the lab suggest that the over-expression of this long non-coding RNA may make the cells more likely to migrate and invade other parts of the body.

Some of the strongest data confirming that long non-coding RNAs are actively involved in cancer, rather than just carried along for the ride, come from prostate cancer. When prostate cancer begins to develop, its growth depends on the male hormone, testosterone. Testosterone binds to a receptor and this leads to activation of various genes that promote cell proliferation. Testosterone binding to its receptor is like you putting your foot down on the accelerator pedal of your car. Prostate cancer is initially treated using drugs that stop the hormone binding to its receptor. This is like having something between your foot and the accelerator pedal, so that you can't press down on it to make the car go faster.

But over time, the cancer cell frequently finds a way around this. The hormone receptor finds ways of activating genes irrespective of whether there is testosterone around or not. It's as if someone has put a bag of sugar on top of the accelerator. The pedal is always pressed down and speeding up the car, even if you have your feet on the dashboard. Two long non-coding RNAs that are highly over-expressed in aggressive prostate cancer have been shown to play a critical role in this process. They assist the receptor, driving gene expression even when there is no hormone around, and accelerating cell proliferation. They play the role of the bag of sugar in the car simile. If expression of these specific long non-coding RNAs is knocked down in cancer models, the tumours show a really dramatic decrease in growth, supporting the critical role of these molecules.[24]

Another long non-coding RNA has also been implicated in prostate cancer. The higher the levels of this long non-coding RNA, the more aggressive the cancer, the shorter the recurrence time after treatment and the greater the risk of death. Knockdown of this long non-coding RNA has a similar protective effect in cancer models to that described above, but in this case the effects do not seem to be due to interactions with the testosterone receptor.[25]

This indicates that long non-coding RNAs may affect cancer progression in different ways, even in one tumour type.

Long RNAs and the brain

It isn't just cancer specialists who are interested in the functions of these molecules. More long non-coding RNAs are expressed in the brain than any other tissue (with the possible exception of the testes).[26] Some have been conserved from birds to humans, with expression patterns that occur in the same regions and at the same developmental stages. These may have conserved functions, perhaps in normal brain development. However, many of the long non-coding RNAs expressed in the brain are specific to humans or primates, and this has led researchers to wonder if they could be responsible, at least in part, for the hugely complex cognitive and behavioural functions found in higher primates.[27]

A long non-coding RNA has been identified that influences how the cells in the brain form connections with each other.[28] Another long non-coding RNA, which has evolved since we diverged from the other great apes, may be involved in regulating a gene that is required for the unique developmental processes that generate the human cortex.[29]

The examples above all suggest that long non-coding RNAs play beneficial roles in the brain. But they may also be implicated in pathology as well as in health. Alzheimer's disease is the devastating dementia which is usually associated with ageing. Because the human population is generally living longer, Alzheimer's disease is becoming increasingly common. The World Health Organization estimates that over 35 million people worldwide are suffering from dementia, and that this number will double by 2030.[30] There is no cure, and even the drugs that are available, which slow down the clinical progression, don't stop it altogether, let alone reverse it. The emotional and economic costs of this condition are enormous, but progress in treating it is horribly slow. This is partly because our understanding of what exactly is going wrong in the brain cells of sufferers is still poor.

We are fairly confident that we know that at least one important step in the process is the production of insoluble plaques in the brain, which can be detected at autopsy. These are made of mis-folded proteins, one of the most important of which is called beta-amyloid. This is generated when an enzyme called BACE1 slices up a larger protein. A long non-coding RNA is produced from the same place in the genes as BACE1, but from the opposite DNA strand, rather like the relationship between Xist and Tsix.

The long non-coding RNA and the standard BACE1 messenger RNA bind to each other. This makes the BACE1 messenger RNA more stable so it stays in the cell for longer. Because it stays around for longer, the cell can generate more copies of the BACE1 protein. This leads to increased production of the beta-amyloid that is essential for the formation of the plaques.[31]

It's been reported that the levels of this long non-coding RNA are increased in the brains of patients with Alzheimer's disease, but it's difficult to interpret these data. This could just be a consequence of increased expression in that region generally. Remember the earlier analogy – the more you chop up logs, the more sawdust you create. But researchers managed to find a way of specifically decreasing the expression of just the long non-coding RNA in a mouse model which frequently develops Alzheimer's pathology. The knockdown of the long non-coding RNA resulted in decreased BACE1 protein and fewer beta-amyloid plaques. This supports the idea that the long non-coding RNA may play a causative role in this devastating disease.[32]

It's not just the central nervous system that can be influenced by long non-coding RNAs. Neuropathic pain is a condition in which the sufferer feels pain, even when there is no physical stimulus. It's caused by abnormal electrical activity in the nerves that conduct signals from the periphery of the body into the central nervous system (brain and spinal cord). It can be a very distressing condition for sufferers, and normal painkillers such as aspirin or paracetamol don't really help. It's often not clear why the nerves are behaving abnormally. Recent work has suggested that in some cases it may be due to increased levels of a long non-coding RNA changing the expression levels of one of the electrical channels. It does this by binding to the messenger RNA molecule that codes for the channel, altering its stability and hence changing the amount of protein produced.[33]

The types of conditions in which it's been claimed long non-coding RNAs play a role is growing all the time.[34] But the controversy remains over how functional and important these long non-coding RNAs really are. Can they really be as important as proteins? Perhaps on an individual basis the answer is usually 'no' unless we are dealing with a molecule as unequivocally vital as Xist. But thinking about the impact in terms of single long non-coding RNAs may be missing the point.

A recent commentary suggested that 'A distinct possibility is that many of the long transcripts are, at best, nudgers and tweakers of genome management, rather than switches per se.'[35] But the greatest complexity and options for flexibility come not from on/off or black/white but from subtle changes in sound levels or from multiple shades of grey. Biologically, we may owe an awful lot to our nudges and tweaks.

9. Adding Colour to the Dark Matter

In biology, the question *What does something do?* is almost always followed by the question *How does it do it?*. We know what long non-coding RNAs are, and we know at least some of what they do – they regulate gene expression. So the perfectly reasonable question is, how do they do that?

There's not going to be one answer to that question. There are thousands and thousands of long non-coding RNAs produced from the human genome, and they almost certainly won't all work the same way. But certain themes are starting to emerge.

One of the most important themes relates to a feature we have already encountered, in Chapter 6, on centromeres and their role in cell division. If we look back to Figure 6.3 (*page 70*) we can remind ourselves that the DNA in our cells is wrapped around bundles of eight histone proteins. So far we have referred to these as packaging proteins, but they actually play far more complex roles than that. Our cells can amend the histone proteins, or the DNA itself. They do this by adding small chemical groups to them. These chemical additions don't change the sequence of a gene. The gene will still code for the same RNA molecule, and the same protein (if it is a protein-coding gene). But the modifications alter the likelihood that a specific gene will be expressed. The modifications are able to do this because they in turn act as binding sites for other proteins. The modifications are the first attachment sites for the build-up of large complexes of proteins that ultimately either switch a gene off or on.

These changes to DNA and its associated proteins are known as epigenetic modifications.[1] *Epi* comes from Greek and means 'at', 'on', 'in addition to', 'as well as'. These modifications are present in addition to the genetic sequence. The easiest modification to understand is the one that is deposited on DNA itself. By far the most common modification to DNA happens when a C base is followed by a G base. This sequence is called CpG, and enzymes in the cell are able to add a modification here. A chemical group called methyl can be added to the C. Methyl is formed of just one carbon atom and three hydrogen atoms and it's very small. Sticking one of these on a C base is like sticking a clover leaf on the side of a sunflower bloom.

If there are a lot of CpG motifs in a stretch of DNA, there are lots of sites where the methyl group can be added epigenetically. This attracts proteins that repress expression of that gene. In extreme cases, where there are lots of CpG motifs in close proximity, DNA methylation can have an exceptionally profound effect. Essentially, the DNA changes its shape and the gene is completely switched off. Remarkably, it can be switched off not just in that cell, but in all the daughter cells that are created when it divides. In non-dividing cells, such as the neurons in our brains, these patterns of DNA methylation may be established while we are in the womb. Many of them will still be in place 100 years later, if we make it that far.

The realisation that DNA methylation could switch genes off more or less permanently during the lifetime of an individual caused great excitement. This was because it finally gave scientists a mechanism to probe something that had been puzzling everyone for decades. Essentially we have known for a long time that not everything can be explained by genetics. We know this because there are a lot of situations where two things are genetically identical and yet the two things are different. When a caterpillar pupates and then turns into a butterfly, it continues to use the same genome. Genetically identical mice, reared under completely standard laboratory conditions, aren't all the same weight.

You and I, dear readers, are masterpieces of epigenetics. The 50–70 trillion cells in a human body pretty much all contain exactly the same genetic code.* Whether they are salt-secreting cells in our sweat glands, the skin cells on our eyelids or the cells that produce the shock-absorbing cartilage in our knees, they all contain exactly the same DNA. They just use the information in those genes in different ways, depending on the tissue. For instance, the neurons in the brain express the receptors for neurotransmitters but switch off the genes for haemoglobin, the pigment that carries oxygen in our red blood cells.

These are all examples of situations we have referred to for decades as epigenetic phenomena. Yes, exactly the same word as for the modifications, and it makes sense. These are all situations where something else is happening in addition to, or as well as, the genetic code.

The discovery of DNA methylation finally gave us a mechanism to understand how epigenetic phenomena happen. In a neuron, the genes responsible for producing haemoglobin become heavily methylated and are switched off. They stay switched off through life. In the cells that give rise to red blood cells, however, these genes are not methylated and haemoglobin is created. But the genes that code for neurotransmitter receptors are switched off using this epigenetic mechanism in these cells.

DNA methylation is pretty stable. It's surprisingly difficult to remove this modification. This is a good thing if your cells need to keep certain genes switched off for long periods. But often our cells need to respond to short-term changes in their environment, if we

* The exceptions are the cells of the immune system that fight off specific infections. Unusually, these cells rearrange some of their genes to create different combinations of antibodies and receptors, able to respond to a vast range of foreign proteins.

drink alcohol or are stressed out by a job interview, for example. Here they turn to a second system. They add modifications to the histone proteins adjacent to genes. Changing the histone modifications can turn genes off, but because these modifications are relatively easy to remove, the cell has the option of turning the genes back on fairly quickly if it needs to. The histone modifications can also be used to modulate the expression of a gene – turn it on a little, quite a bit, quite a lot, a heck of a lot and so on. At a simplistic level we can think of DNA methylation as the on/off switch and histone modifications as the volume control.

The reason histone modifications can act as the fine-tuning mechanism for gene expression is because there are lots of different ones. If DNA is black-to-white with perhaps a few shades of grey depending on the level of methylation, histone modifications are glorious technicolour. There are multiple amino acids that can be modified on histone proteins, and there are at least 60 different chemical groups that can be added to the various amino acids. That creates an extraordinary degree of complexity because at different genes, or the same gene in different cell types, there are thousands of possible combinations of histone modifications. These will be interpreted by the cell in different ways, because they will attract different complexes of proteins that control the gene expression levels and patterns. Some combinations will drive up gene expression, others will drive it down.

Finding a place on the genome

But for years we were faced with a puzzle. The enzymes that add modifications to histone proteins are blind to DNA sequence. They don't bind DNA and they can't distinguish one DNA sequence from another. And yet, in the presence of a relevant stimulus, whatever that might be, the enzymes were very precise in how they modified specific histones. They would add (or remove) modifications at the histones positioned at relevant genes, but ignore nearby histones associated with irrelevant genes.

It's now starting to look as if one of the roles of long non-coding RNAs is to act as a kind of molecular Blu-Tack, attracting histone-modifying enzymes into the vicinity of selected genes. One of the pieces of evidence that this might be the case came from the work analysing the effects of certain long non-coding RNAs in human ES (embryonic stem) cells that was presented in Chapter 8. The researchers showed that about a third of the long non-coding RNAs they examined bound to complexes of proteins that included histone-modifying enzymes. To examine if this binding of long non-coding RNAs to the proteins had any functional consequences, they knocked down expression of the histone-modifying enzyme in the complex. In almost half the cases, the effects on the cell and on gene expression were the same as if they knocked down the long non-coding RNA itself. This suggested that the long non-coding RNA and the histone-modifying enzymes really were working together in the cell.[2]

Many of the investigations of this cross-talk between the long non-coding RNA and epigenetics systems have focused on a specific epigenetic enzyme. This enzyme deposits a specific histone modification that is strongly associated with switching off genes. We can refer to this enzyme as the major repressor.* This has been shown to interact with lots of different long non-coding RNAs.

The long non-coding RNA from a gene targets the major repressor to that gene. The major repressor enzyme then creates repressive modifications on the histones, driving down expression

* The name for this major repressor enzyme is EZH2. It is responsible for adding three methyl molecules to an amino acid called lysine at position 27 on histone H3. The technical nomenclature for this modification is H3K27me3 and it is the best-characterised repressive mark in epigenetics outside of DNA methylation.

of the genes. The repressive modifications attract other proteins, which bind and repress the gene even further.

This control by the major repressor epigenetic enzyme is frequently used to control genes that code for other epigenetic enzymes. Often, these will be genes that have the opposite effect to the major repressor, i.e. they tend to turn genes on. The overall effect is that the major repressor has a strong influence on general patterns of gene expression.[3] It represses genes directly, but also indirectly by preventing expression of epigenetic enzymes that normally switch other genes on. An epigenetic double-punch.

Usually this is a completely normal part of the control of gene expression that happens in our cells, and the system is doing exactly what it's supposed to, making sure that all the complex cellular pathways run in an integrated fashion. But if one part of the complex interaction between long non-coding RNAs and the epigenetic machinery goes out of kilter, problems may develop.

Unfortunately, this seems to be exactly what is happening in some cancers. The major repressor is over-expressed in certain cancers, such as subsets of prostate[4] and breast[5] cancer, and this over-expression is associated with poor prognosis. In certain types of blood cell cancer, the major repressor has mutated, making it abnormally active.[6] The outcome in each case appears to be that the 'wrong' genes are repressed. This creates an imbalance where proteins that drive the cell into proliferation outrun those that usually act as a brake, promoting a cancerous state. Drugs that inhibit the activity of the major repressor are in early clinical trials.[7]

The major repressor works as part of a large complex of proteins*, and various long non-coding RNAs have been shown to be

* This complex is known as Polycomb Response Complex 2 or PRC2. The activity of PRC2 is closely coordinated with that of another repressive complex called PRC1. PRC2 usually establishes the first repressive modifications at a genomic region and PRC1 follows on with additional modifications that stabilise the repressive state.

associated with this complex, suggesting there may be multiple ways of targeting the repressive modifications, depending on the cell type and its behaviour. In Chapter 8 we met a long non-coding RNA whose over-expression drives prostate cancer (*see page 108*). It has been shown to bind to the major repressor and direct it to certain genes, including ones that normally hold back cell proliferation.[8] This finding reinforces the concept that there is a delicate balance of long non-coding RNAs and epigenetic modifiers and that disturbing the equilibrium may be dangerous for a cell or an individual. So do similar data around binding of the long non-coding RNA that is involved in skeletal deformities and a range of cancers, which we encountered in the same chapter (*see pages 106, 108*). It binds to the complex containing the major repressor, and simultaneously to another epigenetic enzyme that can deposit an additional repressing modification.[9]

One of the features implicit in the above explanation is that the long non-coding RNA is transcribed at or near the gene whose histones will be targeted by the major repressor or by other epigenetic enzymes. Although it's difficult to investigate this, the existing data suggest that this is indeed the case. The major repressor can bind to all sorts of long non-coding RNA molecules. The complex containing the major repressor can recognise different types of histone modifications, depending on the components of the complex. These components can vary from cell to cell. As they 'scan' the nearby histones, the complexes can recognise various modification patterns and reinforce these by adding the major repressive modifications. Alternatively, if the region is very rich in modifications that lead to gene expression, the complex may be inhibited and the major repressor will leave the histones alone. This is another of those scenarios where it is a disadvantage to think in purely linear terms, of what came first. Instead, patterns are often maintained or created as a consequence of the histone modification combinations that are already present on the genome.[10,11]

This also seems to hold for the opposite effect, where active regions remain active. Long non-coding RNAs have been reported to be expressed from regions where protein-coding genes are switched on. These long non-coding RNAs stay moored to the genome region where they are produced, possibly by forming a third strand to accompany the double helix of DNA. These long non-coding RNAs bind to the enzymes that place methyl modifications on DNA and stop them doing their job. This keeps the genes in that region in an active state.[12]

If you're inactive, you stay inactive

Xist, which is critical for switching off expression from one of the X chromosomes in a female cell, was one of the first functional long non-coding RNAs to be identified. Perhaps it's no surprise that it's the one whose cross-talk with the epigenetic system has been shown most clearly. As Xist spreads along the X chromosome it attracts other proteins. Many of these are epigenetic enzymes that add chemical modifications to either the DNA or the histone protein. They include the major repressor of histones, and also the enzymes that add methyl groups to DNA.[13] The epigenetic modifications they produce strengthen the shutdown of genes and ultimately lead to hyper-compaction of the inactive X chromosome, and the formation of the Barr Body that we encountered in Chapter 7 (*see page 84*).

It may seem puzzling that the epigenetic modifications always get re-established on the correct X chromosome after cell division. It may be easier to imagine this using a physical example. You have two wooden baseball bats, and you coat one of them with magnetic paint, which represents Xist. After the paint has dried you drop both bats into a tub containing little iron discs. One side of each disc is coated with hooked Velcro. The discs represent the epigenetic proteins that bind to the Xist-coated chromosome. These discs will stick to the bat that has a magnetic covering, but not to the other one. After that, you drop each bat into a tub containing pretty fabric flowers, each backed with a piece of looped Velcro. These represent the modifications. Clearly, the flowers will only stick to the bat that was originally coated with magnetic paint, even though they aren't magnetic themselves.

You could even take this slightly bizarre thought experiment further. Even if you take the flowers off the bat, if you drop it into another tub containing Velcroed blooms, it will be covered again. You could even take off the little iron discs, and as long as you put the bat back into the first and second tubs, it will get covered in flowers again.

In fact, the only way in which you can prevent the bat being covered in flowers when you drop it into the two tubs is to remove the magnetic paint. This is essentially what happens when women make eggs. The inactivating marks are all removed from the X chromosomes and all the daughter cells, i.e. all the eggs are 'fresh' in the sense that they won't pass on inactivation to their offspring. The magnetic paint has to be applied anew to one of the two X chromosomes during early development.

Keeping the ancient aliens quiet

Long non-coding RNAs clearly interact with and help regulate the function of epigenetic proteins. But it would be a mistake to think this is the only way in which junk talks to the epigenetic system. Far from it. We saw in Chapter 4 that the human genome has been invaded by vast numbers of repetitive DNA elements and how important it is that these are kept switched off. Some researchers have gone so far as to speculate that epigenetic control of gene expression may originally have evolved to keep certain junk regions under control.[14] It was only later that the epigenetic system struck out into new territory of regulating normal endogenous genes.

A really striking example of the interplay between junk DNA, epigenetics and the final appearance and behaviour of a mammal can be found in a mouse strain called the Agouti viable yellow mouse. All the mice in this strain are genetically identical, but they can look very different. Some are fat and yellow, some are thin and brown, and others are somewhere in between. The differences in their appearance are due to variable epigenetic regulation of a junk DNA region. In these mice, a repetitive DNA element has become inserted into the genome in front of a particular gene. The DNA element can undergo varying and random degrees of methylation. The heavier the methylation, the more the activity of the repetitive DNA element is repressed, and this affects the expression of the nearby gene.[15] It's the expression levels of the nearby gene that ultimately determine how fat and how yellow the mouse will be. This is summarised in Figure 9.1.

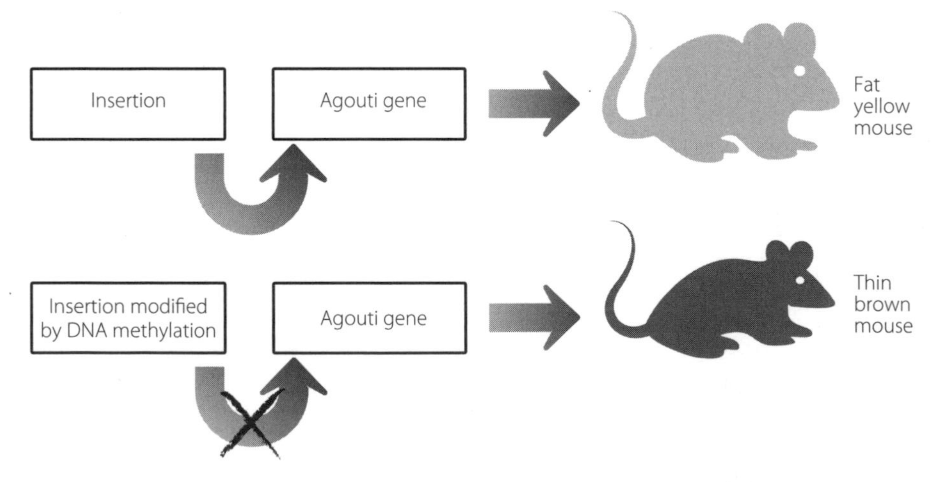

Figure 9.1 In the top panel an insertion drives expression of the *Agouti* gene, leading to a fat yellow mouse. In the bottom panel the insertion has been modified through DNA methylation. The insertion can no longer drive expression of the *Agouti* gene, and the mouse is brown and skinny.

Epigenetics and expansions

The cross-talk between epigenetics and junk DNA is also responsible for the impact of certain genetic mutations. The classic example is Fragile X syndrome, which was described in Chapters 1 and 2. The mutation that causes this condition is the expansion of the CCG triplet repeat, sometimes involving thousands of copies. This repeat contains a C followed by a G – the CpG formation introduced earlier in this chapter as the target sequence for DNA methylation. When this junk repeat sequence gets extremely large, it becomes irresistible to the enzymes and proteins that add a methyl group to the motif. This leads to very heavy methylation of the repeat, which in turn attracts all the proteins that repress gene expression, and even changes in the structure of the DNA itself. The end result is that it is impossible for the cell to express the Fragile X protein, and the consequence of this interaction between junk DNA and epigenetics is a lifetime of learning disability and social disadvantage.

10. Why Parents Love Junk

One of the first Bible stories learnt by children raised in the Judaeo-Christian faiths is the creation tale from Genesis. In this story, God creates the earth and the heavens and all that is in them, and finally he creates Adam and Eve. After that, peopling the earth is down to those two and their descendants, with no further divine intervention apart from the obvious exception in the Christian tradition at the start of the New Testament.

The strong grip of the Adam and Eve story perhaps drives or possibly reflects our ingrained acceptance of a simple piece of biology. To create a child you need a man and a woman. It's not possible, biologically speaking, for a new child to be generated by two men, two women or a woman on her own.

This biological certainty seems so obvious that we virtually never think to question it. But we should, because sometimes the most extraordinary biology lies hidden in the most apparently mundane of assumptions. We should also question it because humans, like all other mammals who bear live young, are in the only class of the animal kingdom where there is never a virgin birth. The mammalian egg needs to be fertilised by a sperm in order to create a new individual. In every other class, there are examples of females who can give birth to live young without ever having mated. It's not just restricted to the lower classes such as insects. Certain species of fish, amphibians, reptiles, and even some birds can do this. Mammals can't, suggesting that this restriction on virgin birth arose relatively recently, following the separation of the mammalian and reptilian lines over 300 million years ago.

We could speculate that this inability in mammals is more of an issue of delivery than fundamental biology. Perhaps two mammalian eggs can't fuse, so they can't create the zygote that can give rise to all other cells. As a consequence, mammalian reproduction would need a male donor because only a sperm can penetrate an egg and deliver its payload of DNA. It's certainly true that mammalian eggs won't normally fuse but this isn't really the explanation. No, the explanation is far more interesting than this, and was demonstrated in a set of elegant experiments in the mid-1980s, using mice as the model system.

Researchers extracted mouse eggs that had been fertilised and took out the nucleus. They reconstituted the eggs using nuclei from eggs or sperm and implanted them back into the uterus of receptive female mice. The results are shown in Figure 10.1.

Live mice were only born after the eggs were reconstituted with a nucleus from both an egg and a sperm. If two sperm nuclei were used, or two egg nuclei, the embryos developed for a little while but then couldn't develop any further. In genetic terms this is really peculiar. In all three experimental systems, the reconstituted egg contained the correct amount of DNA. In terms of DNA sequence there is no actual difference between the DNA from the sperm and the DNA from the egg, particularly because the experiments were designed so that the egg and the sperm each contributed an X chromosome. This created a strange paradox. In all three experimental situations, the DNA sequences involved were exactly the same. But live young only developed if those DNA sequences were donated by a male and a female.[1]

We can be very confident that this requirement for both an egg and a sperm isn't something restricted to mice, because of a human condition called a hydatidiform mole. A woman may appear to be pregnant, gaining weight and frequently suffering from extreme

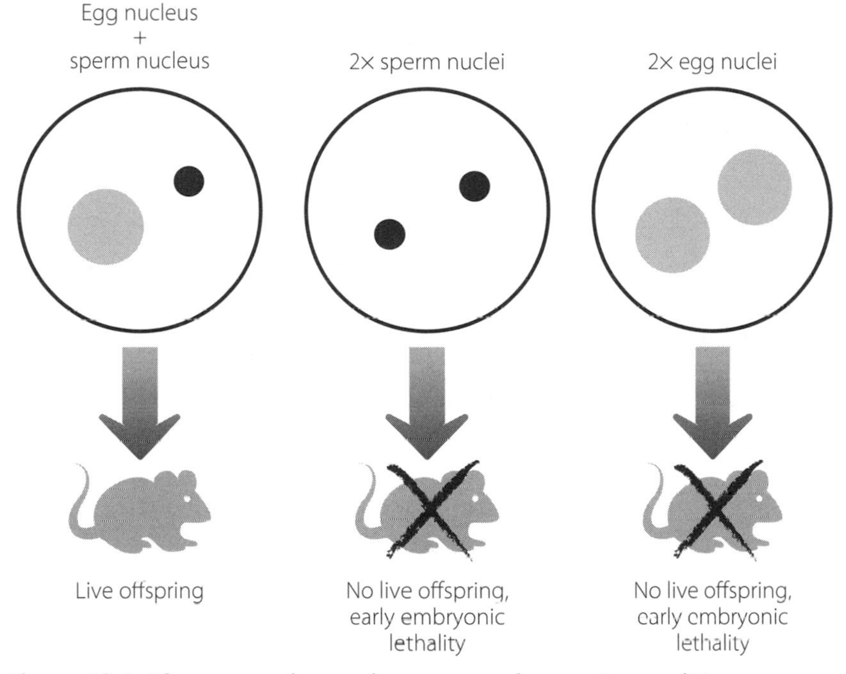

Figure 10.1 If an egg nucleus and a sperm nucleus are inserted into an empty egg which has lost its own nucleus, a live mouse is generated. If two egg nuclei or two sperm nuclei are used, the resulting embryos fail to develop properly. The same genetic information is present in all three scenarios.

morning sickness. But a scan will show the presence of an enlarged abnormal placenta, full of fluid-filled lumps, and no embryo. This is a hydatidiform mole and it is detected in about 1 in 1,200 pregnancies, although in some Asian populations this figure can reach 1 in 200. The structure will abort spontaneously at about four to five months post-fertilisation, although in societies with good prenatal care clinicians will remove it earlier to prevent the development of potentially dangerous tumours.

Genetic analyses of the abnormal placenta have been very informative. They show that in most cases, hydatidiform moles arise when an egg that for some reason has no nucleus is penetrated by a sperm. The 23 chromosomes in the sperm are copied, to create the normal human chromosome number of 46. In about a fifth of cases the mole is formed when two sperm penetrate one of the unusual nucleus-free eggs simultaneously, again generating the correct number of chromosomes. Just like the mouse experiments, the hydatidiform mole contains the correct number of chromosomes but they derive just from one parent, and this again leads to a severe failure in developmental pathways.

The clinical situation and the mouse experiments demonstrated something really fundamental. They showed that the gametes (eggs and sperm) contribute other information in addition to the genetic code. The findings simply can't be explained by the DNA amount or sequence. At a phenomenon level, this is an example of epigenetics. We now know that at the molecular level the phenomenon is caused by the interaction of the epigenetic system and junk DNA.

Remembering where DNA comes from

Scientists have discovered that certain regions of DNA carry epigenetic modifications that indicate 'I'm from mother' or 'I'm from father'. This is known as a parent-of-origin effect. In these regions of the genome, normal development is critically dependent on inheriting one copy of a specific gene (or genes) maternally and the other paternally.

The epigenetic modifications don't just act as pieces of blue or pink genetic decoration indicating who gave you which copy of a gene. The modifications control the expression of specific genes, so that in each matching pair one will be turned on (for instance, the one you inherited from your father) but the one inherited from the other parent (in this case your mother) will be switched off. This system is known as imprinting, because the genes have been imprinted with information about their origin.

Normally, the fact that cells express two copies of a

protein-coding gene gives that cell a kind of insurance policy. Even if one of the copies suffers a mutation or perhaps is inappropriately silenced through abnormal epigenetic modifications, the cell has another copy to fall back on. But if the cell has switched off one of the copies through imprinting, this leaves it more vulnerable to the effects of a random shutdown of the other copy. The fact that for some genes the cell is willing to take this risk tells us that there must be substantial benefits to imprinting that outweigh this disadvantage.

It's no accident that this system has only arisen in mammals. Female mammals make an extraordinarily large investment in the development of their offspring. They keep them inside their body, sharing nutrients with them via the placenta. Now, there are plenty of examples in other classes where a female invests in her young. Think of birds incubating their eggs, or crocodiles building elaborate nest piles and carefully regulating the temperature. But in no other class does the female actually nourish the developing embryos so dramatically.

But for good evolutionary reasons, there is a limit to this degree of maternal commitment. In terms of passing on her genes successfully, the female mammal would prefer to have more than one shot on goal. It's possible that there may be other potential mates who are fitter (in evolutionary terms) than the one whose offspring she is carrying. So although she invests a lot in each pregnancy, it makes sense for the female to be able to breed more than once. There is a definite benefit to ensuring that the developing embryo or embryos gain enough nutrition from her that they have an excellent chance of surviving and reproducing themselves. But it doesn't make sense to divert such a large amount of nutrition to the embryo that the mother ends up losing so much condition that she doesn't survive or is subsequently infertile.

But the same isn't really true for the male. It doesn't really matter to him if his offspring draw so much nutrient from their mother that she never reproduces again. In evolutionary terms, all he wants is for his offspring to be as well-nourished and strong as possible, so that they have the greatest chance of reaching sexual maturity and passing on his genes. He is likely to breed with other females, as relatively few mammals mate for life.

Female mammals can't make decisions about the proportion of nourishment they pass on to the embryos in the uterus. They aren't like birds who can abandon a nest. So evolution has reached an epigenetic stand-off in a nutritional arms race. Imprinting has evolved to balance out the competing demands of the male and female contributions to the genome. At a small number of genes, epigenetic modifications on the DNA inherited from the father set up patterns of gene expression that promote embryo growth. At the same genes, a different pattern of epigenetic modifications on the DNA inherited from the mother has the opposite effect.

During development, the relevant paternal genes often drive expression of a large, efficient placenta, as this is the organ that nourishes the embryos. That's why in the hydatidiform moles, where all the genetic material is from the father, there is an abnormal and very large placenta.

Switching off by switching on

The number of imprinted protein-coding genes is fairly small, about 140 in mice.[2] They occur in clusters of between two and twelve genes and many of these clusters are quite similar to those in the human genome.[3] Perhaps unsurprisingly, the number of imprinted genes is much lower in marsupials where the young are only nourished in utero for a rather short period.[4]

The most critical component in each cluster is a region of junk DNA that controls the expression of the protein-coding genes. This critical component is called the imprinting control element, or ICE. It's a little like lighting a room with twelve light bulbs.

If you want to adjust the level of light in the room, you could use a range of bulbs with different luminosities, and you could have a separate switch for each. But that's a fairly labour-intensive way of controlling the overall light level. Much better to have all twelve bulbs on one circuit and control them simultaneously with either an on/off switch or a dimmer switch if you want a bit more flexibility.

The ICE acts as the central dimmer switch, but there's a slight complication compared with our electrical analogy. The reason why the ICE is important is because it is responsible for driving the expression of a large non-coding RNA molecule. This long non-coding RNA can switch off the expression of the genes in the surrounding cluster. So, essentially, imprinting is critically dependent on two types of junk DNA – ICE regions on the genome, and the long non-coding RNAs the ICEs control. If the long non-coding RNA at a specific cluster is switched on, it switches off expression of the protein-coding genes in that cluster. On the other hand, if the long non-coding RNA driven by the ICE is repressed, the protein-coding genes in the cluster can be activated.

Imprinting critically depends on the junk DNA and its cross-talk with the epigenetic system. The ICE can be epigenetically modified. Expression of the long non-coding RNA is dependent on whether or not the DNA at its ICE is methylated. If the ICE DNA is methylated, this prevents expression of the long non-coding RNA. If the ICE has escaped methylation, the long non-coding RNA will be expressed. Essentially there is a reciprocal arrangement. If the long non-coding RNA is expressed, the genes in the cluster on the same chromosome will be switched off. If the long non-coding RNA is not expressed, the genes in the cluster on the same chromosome will be switched on. The long non-coding RNAs in the imprinted regions are sometimes extraordinarily long, the biggest being a staggering 1 million bases in length.[5]

Unfortunately, we are still a bit sketchy in our understanding of the exact mechanisms that the long non-coding RNAs use to repress the expression of the nearby gene cluster. It certainly seems to involve the epigenetic system again, resulting in the deposition of repressive epigenetic modifications on the protein-coding genes. If key epigenetic genes such as the major repressor that we met in Chapter 9 are knocked out in developing embryos, some of the imprinted genes that would normally be switched off are expressed.[6] It's not just restricted to the major repressor either, as knockout of other epigenetic genes that establish repressive histone modifications has similar effects.[7,8] This demonstrates the importance of the epigenetic system in carrying out the instructions of the long non-coding RNA. It's likely this is because the long non-coding RNA attracts these enzymes to the imprinted cluster, thereby targeting the histone modifications to the protein-coding genes.

Epigenetic modifications are also present at the ICE itself. As we would expect, if the ICE DNA is methylated, the histone modifications are ones which are associated with switching genes off. If the ICE is unmethylated, the histone modifications are those which are associated with switching genes on. The pattern of epigenetic modifications on the ICE is completely consistent across the DNA and histone proteins.[9]

In the imprinting process, the critical determinant is whether or not the junk DNA forming the ICE is methylated or not. There have been suggestions that the methylation of the ICEs evolved when silencing of nearby parasitic elements such as those we met in Chapter 4 spread into neighbouring regions. This may have conferred a fitness advantage, and been selected for in subsequent generations.[10] It's intriguing that in the most primitive mammals, the egg-laying monotremes such as the duck-billed platypus and the echidna, there are uncharacteristically few parasitic elements near the regions where we would expect to find the ICEs in higher mammals.[11]

Resetting the imprint

But how does the methylation pattern become established at the ICE in modern mammals and passed on, given that it is not dependent on differences in DNA sequences between the maternally and paternally derived genomes? How does it get set properly? A woman will inherit imprinted regions from her father in which the ICE is methylated/non-methylated to signify she received this region from her dad. But if she passes this same imprinted region on to her child, this paternal imprint must be removed and replaced with one showing it came from mother.

This seems full of internal contradictions, but it becomes a little easier to understand if we once again visit the world of the musical. Not Oscar Hammerstein this time but Hal David, who was the lyricist who worked for a long time with Burt Bacharach. They wrote the songs for the 1973 flop film musical *Lost Horizon*. One of the songs from this became famous and contains a quite useful concept for us: 'The world is a circle without a beginning and nobody knows where it really ends.' Developmental processes become much easier to visualise if we think of them as never-ending circles rather than in straight lines. The 'put it on–take it off–put it on' cycle in the generation of the imprinted ICE is shown in Figure 10.2. This shows how eggs always pass on a maternal pattern of ICE methylation. A similar process allows sperm always to pass on the reciprocal paternal pattern.

Of course, one of the questions this schema raises is how the developing eggs and sperm identify ICE regions and how they 'know' which should be methylated and which unmethylated. This is an area of very active research and it may be different for each ICE, and between male and female germ cells. Some of it is frankly still a mystery but there are certain features that have been elucidated. We know that in the maternal germline, i.e. the cells that give rise to eggs, the process is critically dependent on the enzymes that can add DNA methylation to previously unmethylated CpG

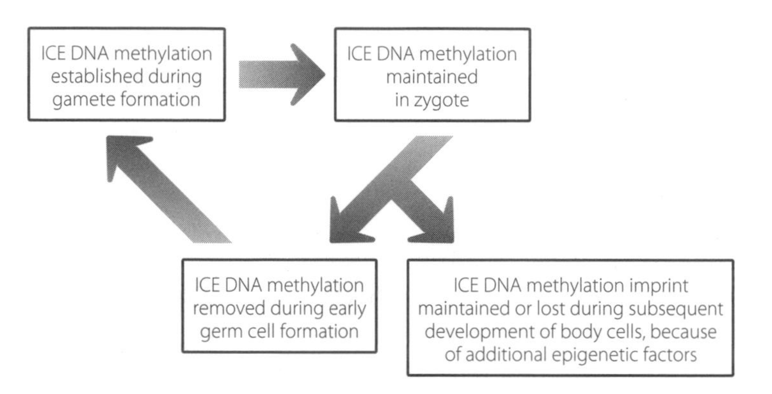

Figure 10.2 Cycles of methylation and demethylation ensure that chromosomes are passed on to children with the correct modifications to indicate parent of origin.

motifs.*[12] After that, the pattern is actively sustained by an enzyme whose job it is to maintain existing methylation patterns.**[13] Other proteins are also likely to be involved in establishing the correct methylation patterns, some of which are likely to be selectively expressed in the developing germ cells.

How do the enzymes in the germ cells recognise the ICE regions among all the other genomic DNA? Again there are gaps in our knowledge, although it has been suggested that certain repeated sequences in these special junk DNA regions may play a role.[14] These are quite poorly conserved at the sequence level between species, but may look more similar when we consider their three-dimensional structures. The cell may have a way of recognising them through their shape, rather than their sequence.[15] This is similar to the findings for long non-coding RNAs we saw in Chapter 8.

Although there are obviously plenty of questions that remain

* The key proteins are called DNMT3A and DNMT3L, the de novo DNA methyltransferases.

** This protein is called DNMT1 and it is known as a maintenance DNA methyltransferase.

about imprinting, we are confident that this is absolutely the reason why both sexes have to contribute to the offspring. In a complex set of breeding experiments using genetically modified mice, researchers showed in 2007 that they could generate viable mice by inserting two egg nuclei into one fertilised egg. The reason they were able to do this was that they artificially altered the pattern of imprinting at two regions in the mouse genome. In one of the egg nuclei, they had created a methylation pattern that looked like the normal paternal pattern, not the maternal one. This fooled the developmental pathways into believing that the genetic material was from a male rather than a female. This demonstrated a particularly strong role for these two imprinted regions in controlling development. It also showed that the only real block to bi-maternal reproduction is the DNA methylation pattern at key genes. It disproved a previous hypothesis that sperm were required because the sperm themselves carried certain necessary accessory factors such as particular proteins or RNA molecules required to kick-start development properly.[16]

Going back to Figure 10.2 we can see that imprinting patterns may change during development. Imprinted control of gene expression seems to be particularly important during development. In mice, for example, most of the 140 or so imprinted genes are only imprinted in the placenta. In adult tissues both or neither copy of the genes may be expressed. This confirms that control of growth during early development was probably the major reason why imprinting evolved. There seems to be almost a geographical reason for this. In the imprinting clusters, the genes nearest the ICE may remain imprinted in all tissues but the ones further from the control centre may only be imprinted in the placenta. Selected cell types in the brain seem to be particularly likely to retain imprinting, although there is no clear consensus on why this would be favoured evolutionarily in most cases. There have been suggestions that the long non-coding RNA produced from the ICE attracts DNA methylation to the nearest genes but attracts histone modifications to the more distant genes in the cluster.[17] Because histone modifications can be more easily altered than DNA methylation, this may provide a mechanism for releasing more distant genes from imprinting as tissues mature.

So, imprinting occurs, and we have insights into at least some of the mechanisms by which this happens. In light of the theory that imprinting has evolved to balance out the competing evolutionary drives of the mother and foetus (and thus indirectly the father), it's not surprising that the majority of protein-coding genes controlled by imprinting are ones involved in foetal growth and infant suckling, along with metabolism.[18] It's also not surprising that when imprinting goes wrong, defects in growth are the commonest symptoms.

When imprinting goes bad

Studies of imprinting disorders really took off in the 1980s, when it was first becoming possible to identify genes associated with inherited diseases. The techniques involved finding families with more than one individual affected by a condition, and then analysing these families to narrow down the region on a chromosome that caused the disease. We can do this pretty easily now because we have the sequence of the normal human genome and access to very cheap sequencing technologies. But back in the 1980s, finding a mutation which caused a disease was akin to being asked to find a specific broken light bulb when all you knew was that it was in a house in America. It took years of work by large teams of scientists to identify the mutations underlying a condition.

A number of groups were looking into a disease called Prader-Willi syndrome. Babies born with Prader-Willi syndrome have a low birth weight and poor suckling responses. Their muscle tone doesn't develop properly until after weaning, so the babies are quite floppy. As the children get older, their appetite becomes completely insatiable and as a consequence they develop early and extreme obesity. The children also suffer from mild mental disability.[19]

A completely different set of researchers was working on a condition with very different symptoms. This is called Angelman syndrome. Children suffering from this condition have small, under-developed heads, severe learning disabilities and are very late at moving on to solid food. The children are prone to outbursts of laughter for no reason, but thankfully the previous appallingly insensitive description of these patients as 'happy puppets' is falling into disuse.[20]

Imagine building a railway across a continent, where one set of workers starts in the east and builds westward, and the other starts in the west and builds eastwards. At first the workers are in completely different territories, but as time goes on they begin to get closer and closer to each other, and eventually there is a point (assuming all has gone well) where they meet, drive in the last spike, shake hands and have a drink. This is pretty much what happened to the researchers investigating Prader-Willi syndrome and Angelman syndrome. The difference, of course, compared with our railway analogy is that the scientists never expected to meet. They thought they were building independent railways, to completely different cities, and yet they each ended up in exactly the same spot as the other.

As the mapping of the chromosomal regions responsible for Prader-Willi syndrome and for Angelman syndrome gathered pace, it became obvious that the two disorders were located in the same region of the genome. At first, the most obvious assumption was that the disorders were caused in genes that were different from each other, but in very close proximity, like two adjacent shops on a street. But eventually it became clear that the disorders were caused by a defect in exactly the same tightly defined region.

Both conditions had the same underlying genetic cause, a loss of a small region on chromosome 15. The parents of the affected children didn't suffer from either disorder and when researchers analysed their chromosomes, they discovered these were completely intact. The loss of the key region of chromosome 15 happened during formation of eggs or sperm.*

It was really bizarre that the deletion of small part of a chromosome could cause two conditions that were so different from each other. But the conundrum began to make more sense once researchers demonstrated that it wasn't just the fact that this small region of chromosome 15 was missing that was important. What mattered was why it was missing. Seventy per cent of children with Prader-Willi syndrome inherited the abnormal chromosome 15 from mutated sperm cells. Seventy per cent of children with Angelman syndrome inherited the abnormal chromosome from mutated egg cells. A little later scientists discovered that 25 per cent of the patients with Prader-Willi syndrome had two perfectly intact chromosomes; nothing was missing. The problem in these patients was that they had inherited both their copies of chromosome 15 from their mother, instead of one from each parent.** In a smaller percentage of Angelman syndrome, the patients had two perfect copies of chromosome 15, but both inherited from their father.

These inheritance patterns make sense only in the context of imprinting, as shown in Figure 10.3. In all the abnormal situations, the cells are lacking an imprinting control region from one parent. This results in abnormal expression levels of the genes that should normally be kept under tight parent-of-origin control, and this leads to pathology including under- or over-growth.

---

* This is known as a de novo mutation, meaning newly arising.

** This is known as uniparental disomy, in this case maternal uniparental disomy.

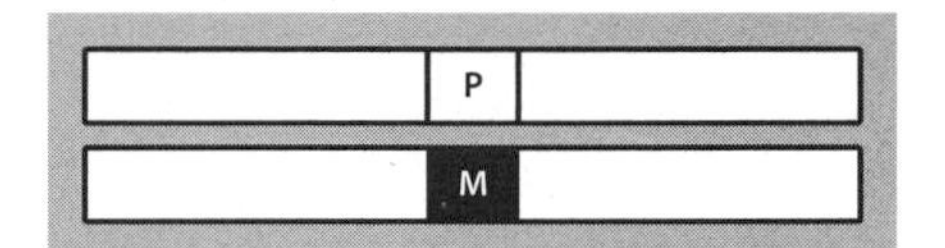

Normal chromosome 15 combination

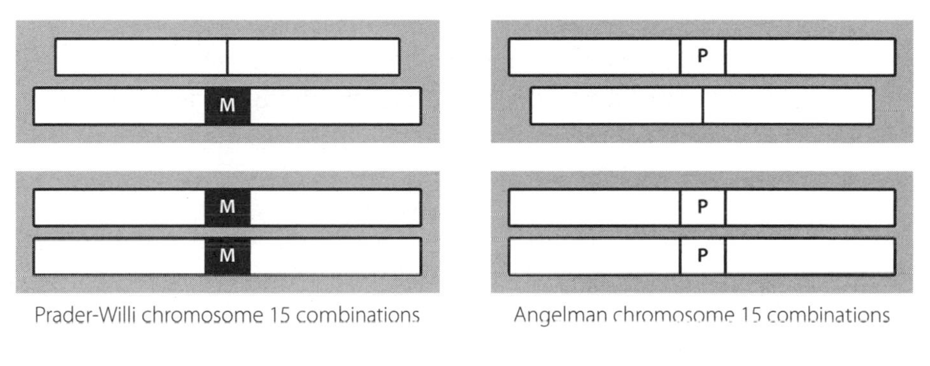

Figure 10.3 Normally, we inherit one copy of chromosome 15 maternally and one paternally. If both copies are inherited maternally, the affected child will have Prader-Willi syndrome. This is also the case if the copy inherited from the father has lost the imprinted region that carries the paternal signature of epigenetic modifications. Essentially, a lack of paternal-specific information leads to Prader-Willi syndrome. Angelman syndrome is caused by a defect in exactly the same region of chromosome 15, but in this case the condition is caused by a lack of maternal-specific information.

Researchers have been able to narrow down the problems that result in these conditions even further, by analysing the genes that are controlled by the imprinting control regions. In about 10 per cent of cases of Angelman syndrome, the patients have inherited all the appropriate DNA from each parent. The problem they have is that there is a mutation in the DNA from their mother. This is not in the ICE but in a gene controlled by the ICE. This is a protein-coding gene, which is normally expressed only from the maternal chromosome. The version on the paternal chromosome is kept silent by imprinting. If the maternally derived gene can't create protein because of a mutation, it means the cell can't produce any of this protein at all, and this leads to pathology.*

The situation in Prader-Willi syndrome is rather more peculiar. A small number of patients have been identified who are lacking just one of the genes that is found in the critical region on chromosome 15. This gene doesn't code for a protein. Instead, it codes for a batch of non-coding RNAs, all of which have similar functions.[21,22,23] These functions are involved in the control of yet another class of RNA molecules that don't code for proteins. It seems to be the absence of this one non-protein-coding gene that is critical for the majority of symptoms in Prader-Willi syndrome.

Consider the implications. A region of junk DNA (the ICE) controls the expression of a piece of junk DNA that encodes a long non-coding RNA. This long non-coding RNA in turn critically regulates the expression of a gene that codes for a batch of non-coding RNAs. And the role of these non-coding RNAs is to regulate other RNAs that don't code for proteins. When we think of it in these terms, it becomes quite difficult to argue that junk DNA has no function.

Prader-Willi and Angelman syndromes are not the only human conditions whereby defects in imprinting lead to abnormalities in growth plus associated problems such as learning disabilities. Another reciprocal pair of conditions are Silver-Russell syndrome,[24] an under-growth condition, and Beckwith-Wiedemann syndrome,[25] which is characterised by over-growth. The two conditions are caused in some patients by parent-of-origin issues at the same region of chromosome 11. It's a particularly complex imprinting locus, with lots of genes involved and more than one ICE.

Similar relationships can be found at other chromosomes.

* This gene is called UBE3A. It adds a molecule called ubiquitin to other proteins, and this leads to degradation of those proteins.

Children who inherit both copies of chromosome 14 from their mother are growth-restricted in the pre- and post-natal periods but later become obese.[26] But if both copies of chromosome 14 are obtained from the father, an abnormally large placenta develops and the child is born with different problems including defects in the abdominal wall.[27,28]

For most of these disorders, there are also rare examples of the condition developing because of epigenetic mistakes. There are small numbers of patients who have inherited the correct DNA from the correct parent. The DNA is not mutated and yet the patients develop an imprinting condition. In these rare cases, the establishment and maintenance of the imprint in the zygote and in early development has usually gone wrong. This can result in an ICE being inappropriately methylated or non-methylated and switched off or on when it shouldn't be. This once again demonstrates the importance of the cross-talk between junk DNA and the epigenetic machinery.

The impact of a dramatic event

In 1978 a little girl called Louise Brown was born. If you had seen Louise Brown you would have thought she was a perfectly ordinary baby. No doubt her parents thought she was the most remarkable baby in the world. What parent doesn't think this about their child? But on this occasion Mr and Mrs Brown could be forgiven for making this claim because they were right. Louise Brown's birth was front-page news all around the world, because she was the first test tube baby.

Her mother's egg had been fertilised by her father's sperm in a dish in a lab and then replaced into her mother's womb. This procedure was used because Louise's mother's fallopian tubes were blocked and she couldn't conceive naturally. The successful birth of Louise Brown opened a new era in treatment of human infertility.

It has been estimated that since that first baby over 5 million children have been born using assisted reproductive technologies.[29]

There have been claims that assisted reproductive technologies may result in higher levels of imprinting disorders, especially Beckwith-Wiedemann, Silver-Russell and Angelman syndromes. The concerns arise because the embryos are being cultured in the laboratory during the critical period when imprinting gets established. It may seem strange that we don't know if there really is a problem or not. Surely with 5 million children to analyse it should be quite straightforward to perform the calculations? But the problem is that imprinting disorders are rare, only occurring naturally at rates of one in several or even tens of thousands. When you are analysing events that are so rare, the statistics can be skewed really easily.

Remember Concorde, one of only two supersonic plane models that ever entered commercial service? For decades it was the safest passenger plane in the world, because there had never been a fatal crash. But following the tragic accident at Paris Charles de Gaulle airport in 2000 in which 109 passengers and crew were killed, it became one of the most unsafe planes, statistically speaking. Of course, this was simply because there were relatively few Concorde flights compared with most airliners and the passenger numbers were small (it was a surprisingly bijou plane inside). Therefore, one event could have a major effect on the statistics if these were calculated in a fairly simplistic fashion.

It's just the same with imprinting disorders. If you would normally expect to see 50 cases for every 5 million children born, how do you interpret it if you detect 55 among the children born via assisted reproductive technologies? Has the medical intervention led to a 10 per cent increase in imprinting disorders, or is this just statistical noise?* We also have to bear in mind that infertility

* The numbers here are random ones, just chosen to demonstrate the point.

itself may lead to a slight increase in imprinting problems, which is simply unmasked by the assisted reproductive techniques. It's possible that sperm or eggs from people with low fertility are more likely to carry imprinting defects, but these only become apparent because they are able to have children thanks to medical technology. In the past, they wouldn't have been able to reproduce so we wouldn't have seen the effects of the imprinting defect.[30] It's one of those confusing situations in biology where what we think we see is possibly distorted because of what's out of sight.

11. Junk with a Mission

It's quite possible that the most wonderful and compelling aspect of biology is its glorious inconsistency. Biological systems have evolved in magnificently creative ways, usurping and repurposing processes for completely new uses wherever possible. It means that almost every time we think a theme is emerging, we find exceptions. And sometimes it can be very difficult to unravel which is the norm and which the exception.

Let's take junk DNA and non-protein-coding RNAs. Based on pretty much everything we have seen so far, it would be perfectly reasonable to develop a hypothesis along the following lines:

When junk DNA encodes a non-protein-coding RNA (junk RNA), the function of the RNA is to act as a kind of scaffold, directing the activity of proteins to particular regions of the genome.

This hypothesis would certainly be consistent with the roles of long non-coding RNAs. They act as the Velcro between epigenetic proteins and DNA or histones. The proteins frequently operate in a complex, and at least one member of the complex is often an enzyme, i.e. a protein that can bring about a chemical reaction. This can be the reaction that adds or removes epigenetic modifications on DNA or histone proteins, or that adds another base to a growing messenger RNA molecule.

In all these situations, the protein is the verb in the molecular sentence. It's the 'doing' or action molecule.

Attractive as this model sounds, it has one unfortunate flaw.

There is a situation where the roles are entirely reversed. In this reversed situation, the proteins are relatively silent, but the junk RNA acts as an enzyme, causing a chemical change to another molecule.

This sounds so peculiar that it is tempting to assume that it's a one-off quirky exception. But if that's the case, it's a really quite remarkable exception because the junk RNA molecules that have this function account for about 80 per cent of the RNA molecules present in a human cell at any given time.[1] We've actually known about these peculiar RNA molecules for decades, making it yet more surprising that we have maintained such a protein-centric vision of our genomic landscape.

The RNA molecules with this odd function are called ribosomal RNA molecules, or rRNA for short. Logically enough, they are mainly found at structures in the cell called ribosomes. These structures are not in the nucleus but in the cytoplasm, which we first encountered in Chapter 2 and which was shown in Figure 2.3 (*see page 16*). The ribosomes are the structures where the information in the messenger RNA molecules is converted into strings of amino acids joined together, creating protein molecules. Using our analogy of the knitting pattern from Chapters 1 and 2, the ribosomes are all the ladies knitting away and turning the information on the printed page into warm socks and gloves for the overseas soldiers.[2]

If analysed by weight, the rRNA makes up about 60 per cent of the structure of a ribosome, and proteins make up the other 40 per cent. The rRNA molecules cluster into two major sub-structures. One contains three types of rRNA and around 50 different proteins. The other sub-structure contains just one type of rRNA and around 30 proteins. The ribosome is sometimes referred to as a macromolecular complex as it is a really big, structured conglomeration of many different components. We can think of it as a large protein-synthesis robot.

When messenger RNA molecules are produced for protein-coding genes, these messenger RNAs are transported out of the nucleus and over to the region of the cell where the ribosome robots are located. Messenger RNA molecules are fed through the ribosome and the genetic instructions carried on the messenger RNA are 'read' by the ribosomes. This results in amino acids being connected together in the correct sequence. It's the ribosomal RNA that carries out the reaction by which an amino acid is joined to its adjacent neighbour. This creates the long, stable protein molecule.

As the messenger RNA is fed through the ribosome, another ribosome may bind at the start of the same message. It too will create protein chains. This is why one messenger RNA molecule can be used as the template for multiple copies of the same protein. The process is shown in Figure 11.1.

The amino acids are brought to the ribosomes by another type of junk RNA called transfer RNA, or tRNA. These are quite small

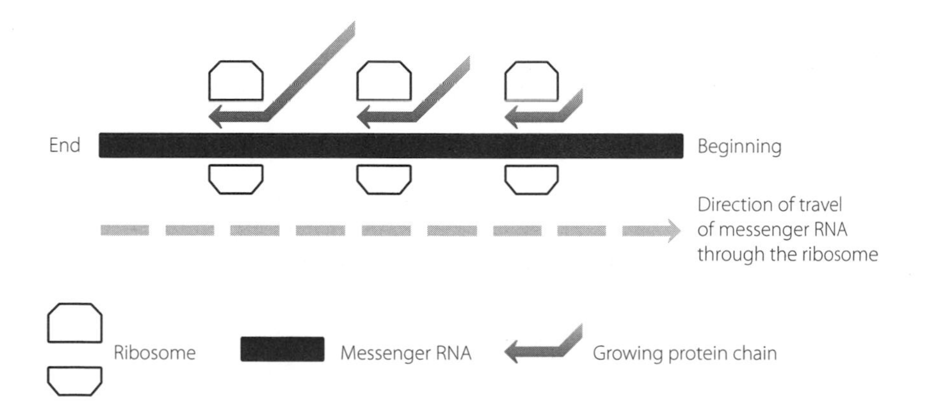

Figure 11.1 A messenger RNA molecule moves through a ribosome, travelling from left to right. The ribosome builds the protein chain. As the beginning of the messenger RNA emerges from the processing ribosome, it can engage another ribosome. As a consequence, there may be multiple ribosomes on a single messenger RNA molecule, all building full-length proteins.

non-coding RNAs, only about 75 to 95 bases in length.[3] But they are able to fold back on themselves creating an intricate three-dimensional structure usually referred to as a clover leaf. A specific amino acid is attached to one end of the tRNA. At the far end, on a loop, is a sequence of three bases. This base triplet can bind to the correct matching sequence on a messenger RNA molecule. It does this by using essentially the same rules as the base pairing in DNA.

The tRNA molecules act as the adapters between the nucleic acid sequence carried on the messenger RNA (and originally the DNA), and the final protein. This ensures that the amino acids are lined up in the right order to create the proper protein. This is shown in Figure 11.2. When two amino acids are held next to each other at the ribosome, the rRNA can carry out a chemical reaction that attaches the end of one amino acid to the beginning of the next and thereby builds the protein chain.

Some of the triplets on the messenger RNA don't have a match to any triplet on a tRNA. These triplets are known as stop signals. When the ribosome reads one of these, it can't fit a tRNA in place

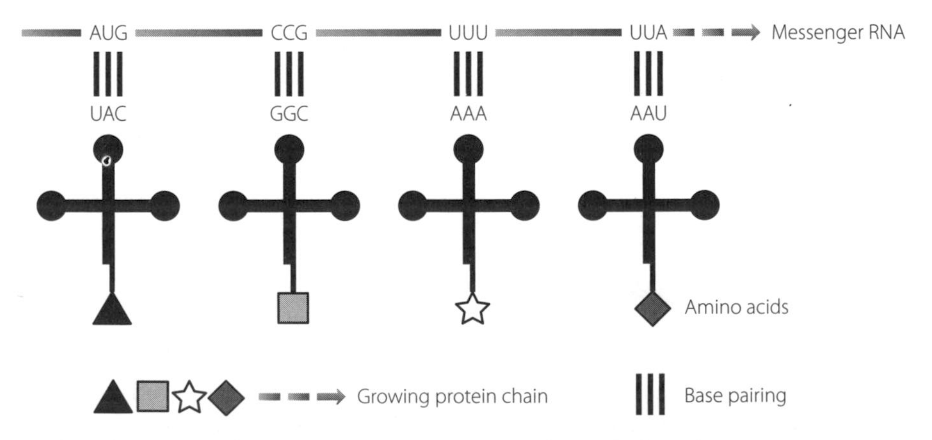

Figure 11.2 As the messenger RNA moves through the ribosomes, transfer RNA molecules bring the appropriate amino acids to the correct position on the chain, using base pairing. The ribosomal RNA machinery joins up adjacent amino acids to create the protein chain.

and the ribosome falls off the messenger RNA and the protein stops growing. These are the roofing LEGO bricks we met in Chapter 7 (*see page 85*). The ribosome then finds another messenger RNA molecule to translate into protein, or could even go back to the start of the first one.

Even though the entire procedure relies on a giant complex of four types of ribosomal RNA and around 80 associated proteins, and is a very sophisticated task, the process of adding new amino acids into a growing protein chain is remarkably fast. It's difficult to measure this accurately in human cells, but in bacteria each ribosome can add amino acids at the rate of about 200 a second. It's probably not as fast as this in human cells, but it will still be about ten times faster than we could possibly stick two bricks together if we were making a LEGO tower. And don't forget that the ribosome isn't sticking together random LEGO bricks. It's as if we had to choose just two out of 20 different types of LEGO bricks (there are 20 different amino acids) and stick them on top of each other in exactly the right order every fraction of a second. Quite a task.

Our cells need to produce millions of protein molecules every second, and so we need our ribosomes to work very efficiently. We also need a lot of ribosomes to meet the demand, up to 10 million robots in a single cell.[4] In order to create enough ribosomes, our cells have accumulated lots of copies of rRNA genes. Instead of being dependent on creating rRNA from the classical situation of one gene inherited from each parent, we inherit about 400 rRNA genes, distributed across five different chromosomes.[5]

One consequence of this vast number of rRNA genes is that we aren't very prone to disorders caused by mutations in these genes. That's because if one copy is mutated we have lots of redundancy. So the chances are that we can make up for the defect from all the normal versions encoding the same rRNA molecule. This isn't true of mutations in the genes coding for the proteins that are also present in the ribosomes. We don't know what many of these do in detail, and some don't seem to be important at all in ribosome function. But there are others where mutations do result in human disorders.

The two best-known examples are called Diamond-Blackfan Anaemia and Treacher-Collins Syndrome. They are caused by inherited mutations in different protein-coding genes. The consequence in both cases is a decrease in the number of ribosomes. But there are clearly subtleties that we don't understand in how this affects cell function, because if the only important factor was this reduced number, we would expect the clinical consequences to be identical. But they aren't. The major symptom in Diamond-Blackfan Anaemia is a defect in the production of red blood cells. The major symptoms in Treacher-Collins Syndrome are malformations of the head and face, leading to problems with breathing, swallowing and hearing.[6]

Because we need a lot of ribosomes and hence a lot of rRNA genes, it's not unreasonable that we also need a lot of tRNA genes to ensure that there are plenty of tRNA molecules to transport the amino acids to the ribosomes. The human genome contains nearly 500 tRNA genes, distributed across almost every chromosome.[7] This brings the same benefits as those described above for multiple copies of rRNA genes.

There's also an odd and intriguing possible overlap between rRNA and imprinting. As described in Chapter 10, there are a small number of patients with Prader-Willi syndrome where the disorder has been localised to a junk DNA region that encodes a batch of non-coding RNAs (*see page 140*). These are called snoRNAs, for small nucleolar RNAs.* These non-coding RNAs migrate to a region of the nucleus called the nucleolus, which is very important

* The ones involved in the process described here are from a specific class called snoRNA C/D box.

in ribosome biology. The nucleolus is the place where the mature ribosomes are assembled, as shown in Figure 11.3.

In the nucleolus, the rRNAs and the proteins are modified and then assembled into mature intact ribosomes which are transported back into the cytoplasm, ready to carry out their functions as protein-creating robots. The snoRNAs are required to make sure that certain modifications take place properly on the rRNA molecules. Just as DNA and histone proteins can be modified by the addition of a methyl group, rRNA molecules can also be methylated. The snoRNAs probably facilitate this by finding regions on the rRNA with which they can form pairs. Once again this is possible because of bonding between complementary bases on the two nucleic acid molecules. Once they bind, the snoRNAs attract enzymes that can add methyl groups to the rRNAs. This may

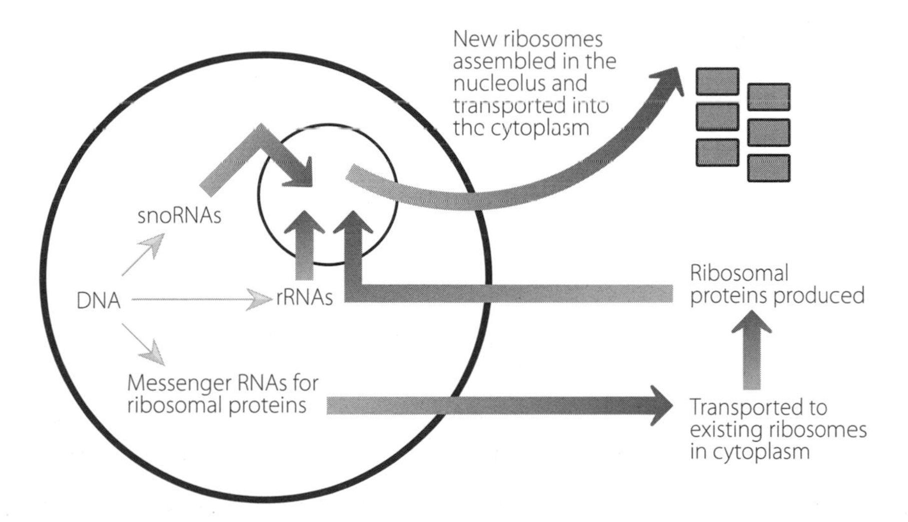

Figure 11.3 Messenger RNA molecules for ribosomal proteins are created in the nucleus and then shipped out to existing ribosomes in the cytoplasm. The new ribosomal proteins are transported back into a specific region in the nucleus. Here they join up with ribosomal RNA molecules to create new ribosomes, which are moved out into the cytoplasm to act.

be similar to the ways in which long non-coding RNAs attract enzymes that modify histones.* It's not altogether clear why these modifications matter to the rRNA, but one suggestion is that they help to stabilise interactions between the rRNAs and the proteins in ribosomes.

Although it's tempting to speculate that the symptoms of Prader-Willi syndrome are caused by problems in the snoRNAs' control of rRNA modifications, this remains just a theory at the moment. The problem is that we now recognise that the snoRNAs can also target lots of other types of RNA molecules, so we can't be sure exactly which process is going wrong in the children with this disorder.

Ribosomes are extremely ancient structures, and can be detected in really primitive organisms. They are even found in bacteria, the tiny single-celled organisms which don't have a nucleus in their cells to separate their DNA from their cytoplasm. Evolutionary biologists often use the DNA sequences of the genes that encode rRNAs to track how species have diverged over time.

Bacteria and higher organisms diverged about 2 billion years ago[8] so although we can still recognise the rRNA genes in our unicellular (very) distant cousins they are really different from ours. This has turned out to be A Good Thing. Some of the most common and successful antibiotics work by inhibiting the bacterial ribosomes.[9] These include tetracycline and erythromycin. These antibiotics disrupt the activity of the bacterial ribosomes, but not human ones. In the West we are so accustomed to antibiotics that we sometimes forget how important they have been, saving literally tens of millions of lives, at a conservative estimate, since they really hit the medical scene in the 1940s. It's odd to think that many of

* The methyltransferase enzyme required for this process is called fibrillarin, which works in a complex with three other proteins and the snoRNA.

these lives have been saved because of variation between species in what purists would consider junk DNA.

We depend on our invaders

It's even odder to think that each one of us has been colonised by organisms that probably developed around the same time our ancestors were diverging from the forebears of modern bacteria. 'Colonised' is really an understatement. Our entire survival and that of every other multicellular organism on this planet from grass to zebras and from whales to worms relies on this colonisation. It's even true of the yeast we depend on for bread and beer.

Billions of years ago the cells of our earliest ancestors were invaded by tiny organisms. At this stage there probably weren't any organisms more than four cells in size and the four cells would have been pretty non-specialist. Instead of warring against each other, these cells and their tiny invaders reached a compromise. Each benefitted from the compromise and so a beautiful friendship, lasting billions of years, was born.

These tiny organisms evolved into critical components of our cells called mitochondria. The mitochondria reside in the cytoplasm and are little power generators. They are the sub-cellular organelles that produce the energy we need to power all of our standard functions. It's the mitochondria that have allowed us to make use of oxygen to create useful energy from food sources. Without them, we would be smelly little four-celled nobodies with hardly enough energy to do anything useful.

One of the reasons we are confident that mitochondria are the descendants of these once free-living organisms is that they have their own genome. It's much smaller than the 'proper' human genome that is found in the nucleus. It is just over 16,500 base pairs in length compared with the 3 billion base pairs of the nuclear genome, and unlike our chromosomes it is circular.

The mitochondrial genome only codes for 37 genes. Remarkably, well over half of these don't code for proteins. Twenty-two of them encode mitochondrial tRNA molecules[10] and two encode mitochondrial rRNA molecules. This allows the mitochondria to produce ribosomes, and to use these to create proteins from the other genes in its DNA.[*,11]

This seems a very risky strategy in evolutionary terms. Mitochondrial function is critical for life and ribosomal function is absolutely critical to mitochondrial function. So why have such an important process with no safety net of extra copies of the ribosomal genes in our power generators?

We can get away with this because mitochondrial DNA isn't inherited in the same way as nuclear DNA. In the nucleus we inherit one set of chromosomes from each parent. But mitochondrial inheritance is different. We only inherit our mitochondria from our mother. This would seem to make for an even riskier scenario because it means if we inherit a mutant mitochondrial gene from our mother, there is no chance of a back-up normal gene from dad.

But there is (of course) a complication. We don't just inherit one mitochondrion from our mother, we inherit hundreds of thousands, maybe even a million. And they aren't all the same genetically, because they haven't all originated from one mitochondrion in a previous cell. Every time a cell divides, the mitochondria also divide and are passed on to daughter cells. Even if some of these mitochondria have developed mutations, there will be plenty of other mitochondria in the cell that are fine.

---

* Mitochondria use lots of other proteins for their biochemical processes, but most of them they import from the cell cytoplasm. The ones that are uniquely encoded in the mitochondria are all involved in a process called the electron transport chain, which takes place within mitochondria themselves. This process is essential for life, as it is how we generate storable usable energy to power our cells.

That's not to say that problems never develop, and many of those that do have been reported to be in the tRNA genes on the mitochondrial DNA. These include conditions with muscle weakness and wasting;[12] hearing loss;[13] hypertension[14] and cardiac problems.[15] But the symptoms may vary a lot from patient to patient, even within the same family. The most likely reason for this is because symptoms may not develop until the percentage of mutant mitochondria in a tissue reaches a threshold. This may not be until relatively late in life, as a consequence of random unequal distribution of 'good' and 'bad' mitochondria when a cell divides.

If all of this hasn't been enough to demonstrate that RNA is not just some poor relation of DNA or an inferior species compared with proteins, consider this. Despite DNA being the poster child for biology, all life on earth may have originated not with DNA but with RNA.

In the beginning was the RNA (possibly)

DNA is a great molecule. It stores a lot of information, and because of its double-stranded nature it's easy to copy and to maintain the sequence stably. But if we try to think back billions of years, to when life began to develop, it's hard to see how it could happen based on a DNA genome.

That's because although DNA is fantastic at storing information, it's no use in terms of creating something from that information, not even another copy of itself. DNA can never function as an enzyme. Because of this, it can't make copies of itself so how could it have been the starting genetic material? It is always reliant on proteins to do its bidding.

But if we look at rRNA, a molecule which has received very little by way of the spotlight even among most scientists, there's a bit of a eureka moment. rRNA contains sequence information but it is also an enzyme. This raises the possibility that RNA could have had a range of enzymatic activities in the past, and this could have led to the evolutionary development of self-sustaining and self-propagating genetic information.

In 2009 researchers published extraordinary work in which they generated such a system. They genetically created two RNA molecules both of which could act as enzymes. When they mixed these molecules in the lab, and gave them the raw materials they needed, including single RNA bases, the two molecules made copies of each other. They used the existing RNA sequences as the templates for the new molecules, creating perfect copies. As long as they were supplied with the necessary raw materials, they made more and more copies. The system became self-sustaining. The researchers went even further by mixing higher numbers of different RNA molecules, each of which had enzymatic activity. When they activated the experiment, they found that two sequences would rapidly outnumber all the others. Essentially, the system was not only self-sustaining, it was also self-selecting because the most efficient pairs of RNA molecules would recreate each other far more rapidly than any of the other pairings.[16] Very recently, scientists have even succeeded in creating a type of enzymatic RNA that will generate copies of itself.[17]

An expression that is still heard in the UK is 'Where there's muck, there's brass', meaning that where there is dirt or rubbish, there's money. Maybe where there's junk, there's life.

12. Switching It On, Turning It Up

With a mere $1,700,000 price tag, the Bugatti Veyron is the world's most expensive production road car. It's hard to be sure what the cheapest car is, although the Dacia Sandero probably has a good claim to this honour, at about 1 per cent of the cost of the Veyron. But both cars have a number of things in common, and one of these is that each needs to be switched on before you can go anywhere. If you don't activate the engine systems, nothing will happen.

Our protein-coding genes are the same. Unless they are activated and copied into messenger RNA, they do nothing. They are simply inert stretches of DNA, just as a Veyron is a stationary hunk of metal and accessories until you hit the ignition. Switching on a gene is dependent on a region of junk DNA called the promoter. There is a promoter at the beginning of every protein-coding gene.

If we think in terms of a traditional car, the promoter is the slot for the ignition key. The key is represented by a complex of proteins that bind to a promoter. These are known as transcription factors. These transcription factors in turn bind the enzyme that creates the messenger RNA copies of the gene. This sequence of events drives the expression of the gene.

It's relatively easy to identify promoters by analysing DNA sequences. Promoters always occur just in front of the protein-coding regions. They also tend to contain particular DNA sequence motifs. This is because transcription factors are a special type of protein that can identify and bind to specific DNA sequences. If we analyse the epigenetic modifications at promoters, we also find consistent patterns emerging. Promoters have particular sets of epigenetic modifications, depending on whether or not the gene is active in a cell. The epigenetic modifications are important regulators of transcription factor binding. Some modifications attract transcription factors and associated enzymes and this results in gene expression. Others prevent the factors from binding and make it really difficult to switch a gene on.

Researchers can copy a promoter and reinsert it elsewhere in the genome, or even into another organism. These kinds of experiments confirmed that promoters usually function immediately in front of a gene. They also showed that the promoter needs to be 'pointing' in the right direction. If you insert a promoter sequence in front of a gene, but the wrong way round, it doesn't work. It would be like inserting a key the wrong way round into the ignition. Promoters are orientation-dependent in their activity.

Promoters can't really tell which gene they are controlling. They switch on the nearest gene, if they are close enough and pointing in the correct direction. This allows researchers to use promoters to drive expression of any gene in which they are interested. That can be very handy experimentally but it can also have a sinister side. In some cancers, the basic molecular problem is that the DNA in the chromosomes becomes mixed up and a promoter starts driving expression of the wrong gene. In the case of cancer, the gene is one that pushes forward the rate at which cells proliferate. The first to be discovered, and probably still the most famous example of this, occurs in the blood cancer known as Burkitt's lymphoma. This is a cancer type we already met briefly, in our discussion of good genes in bad neighbourhoods (*see page 48*). In this condition, a strong promoter on chromosome 14 gets placed upstream of a gene on chromosome 8 that codes for a protein that can really push cell proliferation forwards.*[1] The consequences

---

* The gene codes for a protein called MYC. MYC is also involved in a range of other cancers.

are potentially catastrophic. The white blood cells carrying this rearrangement grow and divide really rapidly, and start to predominate in the blood stream. If detected early in the disease's progression, over half of the patients with this cancer can be cured, although this requires aggressive chemotherapy.[2] For patients with a late diagnosis, decline and death may be appallingly rapid and measured in weeks.

In healthy tissues, different promoters may only be active in certain cell types, usually because they rely on transcription factors that are expressed in some cell types and not others. Promoters also have different strengths. By this we mean that strong promoters switch on genes very aggressively, resulting in lots of copies of messenger RNA from the protein-coding gene. This is what happens in Burkitt's lymphoma. Weak promoters drive much less dramatic levels of gene expression. The strength of the promoter is dependent on multiple factors in mammalian cells, including the DNA sequence but also the transcription factors available, the epigenetic modifications and probably a host of other variables that we haven't yet identified.

Driving a graduated response

Any given promoter in any given cell type drives a relatively constant level of gene expression, at least in experimental systems. Yet gene expression under normal circumstances is not a binary phenomenon. Genes may be expressed to varying degrees. It's analogous to being able to drive a car at any rate from one mile per hour up to its top speed of over 250mph for the Veyron, or rather less than half of that for the Sandero. In cells, this flexibility is dependent on a number of interacting processes including epigenetics. But it is also influenced by another region of junk DNA. This region is known as the enhancer.

Compared with promoters, enhancers are very fuzzy. They are

usually a few hundred base pairs in length but it's almost impossible to identify them simply by analysing the DNA sequence.[3] They are just too variable. The identification of enhancer regions is also made more complex because they aren't necessarily functional all the time. For example, a set of latent enhancers has been identified which only start to regulate gene expression once they themselves have been somehow activated by a stimulus. This showed that enhancers may not be pre-determined in the genome sequence.

An inflammatory response is the first line of defence to an assault on the body, such as a bacterial infection. The cells near the invasion site release chemicals and signalling molecules that create a really hostile environment for the invaders. It's as if triggering a burglar alarm in a house initiated a downpouring of hot, foul-smelling liquid into the room that had been breached.

Scientists studying the inflammatory response were among the first to show that DNA sequences can be co-opted to become enhancers when necessary. In this study, the researchers found that once the inflammatory stimulus was removed, the enhancers didn't revert to being inert. Instead they continued to be enhancers, ready to up-regulate expression of the relevant genes again, if the cells re-encountered the inflammatory stimulus.[4] It's probably not a coincidence that these enhancers are regulating genes involved in the response to foreign invaders. This memory in terms of gene expression can be very advantageous for fighting off an infection as efficiently and swiftly as possible.

Epigenetics and enhancers – cross-talk in action

One way in which genetic regions can maintain a memory even after a stimulus has gone away is via epigenetics. Epigenetic modifications can make a region easier to switch on again, by keeping the region in a fairly de-repressed state. In human terms, it's like a doctor being on call rather than on holiday. In the example above, the researchers demonstrated that certain histone modifications remained at the 'new' enhancers after the inflammatory stimulus was removed, keeping them in a state of readiness.

We are generally starting to make a bit more progress at identifying enhancers by looking at epigenetic modifications, which are independent of the underlying DNA sequences. The modifications can be used as functional markers to show how a specific cell type uses a stretch of DNA. Researchers have also shown that these modifications can change in cancer, creating different patterns of gene expression that may contribute to the cellular alterations that lead to cancer.[5]

But even if we can find an epigenetic signature that indicates we may be looking at an enhancer, we still have another problem. We don't know which protein-coding gene is influenced by a putative enhancer. The only way we can establish this is by disrupting an enhancer using genetic manipulations, and then assessing which genes are directly influenced by this change. This is because enhancer function is different from that of the promoter. Enhancers are not orientation-dependent – they act as enhancers no matter which way they are pointing. The other difference is even more dramatic – enhancers can be a very long way from the protein-coding gene whose expression they are influencing.

There are also far more enhancers than we might expect. A recent comprehensive study looked at the patterns of histone modifications in nearly 150 human cell lines. When they assessed these lines for patterns that looked like enhancers, they found nearly 400,000 candidate enhancer regions.[6] This is far more than required if there was a one-to-one relationship between enhancers and protein-coding genes. It's even too many if we assume that long non-coding RNAs have enhancers.

The enhancers weren't all found in every cell type. This is consistent with a model where the same stretch of DNA can have different functions in different cells, depending on how it is epigenetically modified.

For many years, we had no clear models of how enhancers really work. We now suspect that in many cases they may be critically dependent on another type of junk: the long non-coding RNAs. In fact, specific classes of long non-coding RNAs may be expressed from the enhancers themselves.[7] Many of the long non-coding RNAs we met in Chapter 8 are involved in repressing expression of other genes. But it is now believed that there is also a large class of long non-coding RNAs that enhance gene expression. This was first suggested to be the case for long non-coding RNAs that regulate neighbouring genes. If expression of the long non-coding RNA was increased experimentally, the expression of the neighbouring protein-coding gene also increased. Conversely, if the expression of the long non-coding RNA was knocked down experimentally, the protein-coding gene also showed lower expression.[8]

Further evidence came from analysing the timing patterns for specific long non-coding RNAs and the messenger RNAs they were believed to regulate. Researchers treated cells with a stimulus that they knew caused expression of a specific gene. They found that the enhancing long non-coding RNA was switched on before the messenger RNA from the neighbouring protein-coding gene.[9,10] This is consistent with a model where the long non-coding RNA located in the enhancer is switched on in response to a stimulus, and then in turn helps to switch on expression of the protein-coding gene.

The long non-coding RNA doesn't drive this increase on its own. The process is reliant on the presence of a large complex of proteins. The complex is known as Mediator. The long non-coding RNA binds to the Mediator complex, directing its activity to the neighbouring gene. One of the proteins in the Mediator complex is able to deposit epigenetic modifications on the adjacent protein-coding gene.* This helps to recruit the enzyme that creates the messenger RNA copies which are used as the templates for protein production. There is a consistent relationship between the Mediator complex and the long non-coding RNA. Experimentally generated decreases in expression of either the long non-coding RNA or a member of the complex each lead to decreased expression of the neighbouring gene.[11]

The importance of a physical interaction between long non-coding RNAs and the Mediator complex has been shown by a human genetic condition. This disorder is called Opitz-Kaveggia syndrome. Children born with this condition have learning disabilities, poor muscle tone and disproportionally large heads.[12] The affected children have inherited a mutation in a single gene. This codes for the protein in the Mediator complex that interacts with long non-coding RNA molecules.**

The more that scientists analysed the activity of the Mediator complex, the more interested they became. One of the reasons was that the Mediator complex is responsible for the actions of a group of enhancers with special powers. These are the super-enhancers, and they are particularly important in embryonic stem (ES) cells, the pluripotent cells that have the potential to become any cell type in the human body (*see page 105*).[13]

The super-enhancers are clusters of enhancers all acting together. They are about ten times the size of normal enhancers. Because of this, proteins can bind to the super-enhancers at very high levels, much higher than are found on normal enhancers. This allows the super-enhancers to really ramp up expression of the gene they are regulating. But it's not just the numbers of proteins

* The modification is addition of a phosphate group (one phosphorus atom and four oxygen atoms) to a specific position on histone H3. This modification is usually associated with active genes.

** The Mediator component is MED12.

that bind that interested the researchers. It's the identities of these proteins.

As we saw in Chapter 8, ES cells don't stay pluripotent by chance or passively. In order for ES cells to maintain their potential, they regulate their genes very carefully. Even relatively mild perturbations in gene expression can start to push an ES cell down a pathway that converts it into a specialised cell type. One way of visualising this is to think of a Slinky at the top of a tall flight of stairs. Just the slightest nudge to push it over the edge of the top step is enough to send that Slinky on a very long journey. Perhaps an even better analogy might be a Slinky that is held back from falling down the stairs by a small weight on its trailing end. Remove the weight, and off the Slinky will go.

There are a set of proteins that are absolutely vital for maintaining the pluripotency of ES cells. These are known as master regulators, and they are like the small weight on the trailing end of the Slinky. The master regulators are expressed very highly in ES cells, but at much, much lower levels in specialised cells.

The importance of these proteins was unequivocally demonstrated in 2006. Researchers in Japan expressed a combination of four of these master regulators at very high levels in differentiated cells. Astonishingly, this set in motion a chain of molecular events which culminated in the creation of cells that were almost identical in action to ES cells.[14] This is analogous to a Slinky at the bottom of a flight of stairs moving all the way back up to the top step. The cells created by this route have the potential to be converted into any cell type in the body.* This remarkable work, and the research that followed on from it, has generated enormous excitement because potentially we can create replacement cells to treat a large number of disorders. These range from blindness to type 1 diabetes, and from Parkinson's disease to cardiac failure.

---

* These are known as induced pluripotent stem cells (iPS cells).

Until this new technology was developed, it was extremely difficult to create appropriate cells to treat human conditions. This is because cells from a different individual usually can't be implanted into another person. The immune system will recognise the donated cells as foreign and kill them, as if they were an invading organism. But, as shown in Figure 12.1, we now have the potential to make cells that are a perfect match for the patient.

The 2006 work has spawned an industry potentially worth billions of dollars, and also resulted in one of the fastest awards of a Nobel Prize in Medicine or Physiology ever, just six years after the original publication.[15]

In normal ES cells, some of these master regulator proteins bind at very high densities to the super-enhancers. The super-enhancers themselves are regulating some key genes that maintain the pluripotent state of the cells. The Mediator complex is also present at very high levels in the same locations. Knocking down the expression of a master regulator, or of Mediator, has very similar effects on the expression of these key genes. The expression

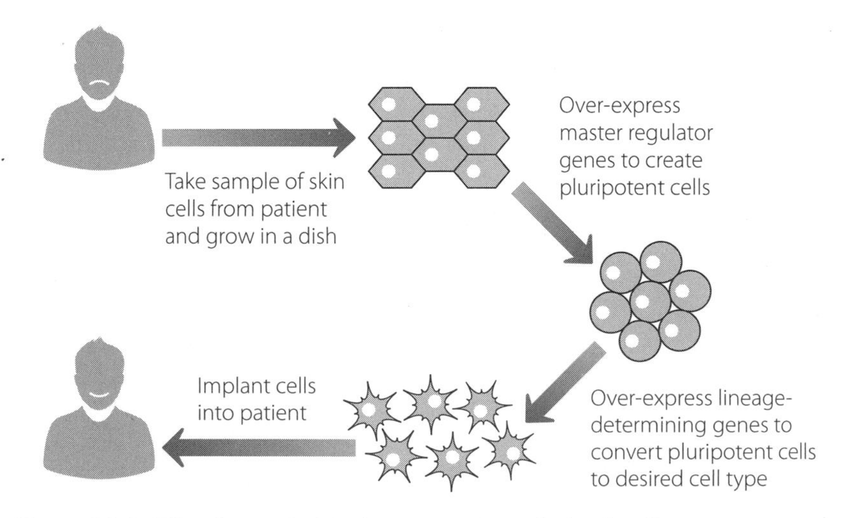

Figure 12.1 The theory behind using patient-derived cells to create therapies tailored for a specific individual.

levels drop, making the ES cells more likely to start differentiating into specialised cell types.

Because the pluripotent state of ES cells is critically dependent on the expression of high levels of the master regulators, it's perhaps unsurprising that the master regulators themselves are controlled by super-enhancers. This creates a positive feedback loop, which is shown in Figure 12.2.

Positive feedback loops are relatively rare in biology, mainly because they can be difficult to get back under control if they start to go wrong. Luckily, the protein-coding genes regulated at super-enhancers are extremely sensitive to small perturbations in binding of master regulators and a number of other factors. This may mean that even a small change in the balance of some of these factors may be enough to interrupt this positive feedback loop, and allow the cells to differentiate rather than remain pluripotent. After all, it doesn't usually take much of a nudge to make a Slinky fall down the stairs.

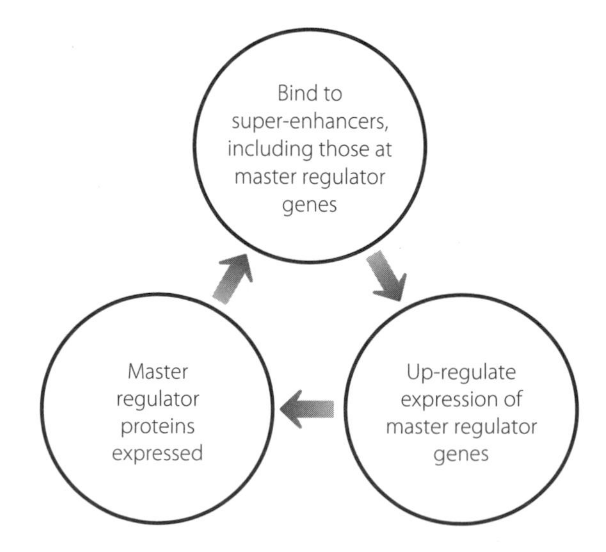

Figure 12.2 The positive feedback loop driving high-level sustained expression of master regulator genes.

Super-enhancers have also been reported in tumour cells, where they are associated with critical genes that drive cell proliferation and cancer progression.[16] One of the genes that is regulated by such a super-enhancer is the same one that we encountered earlier in this chapter, which drives Burkitt's lymphoma. There are also super-enhancers in some normal specialised cells. These bind cell-specific proteins that define cell identity.

Overcoming the distance

Most of the events described so far involve enhancers that are relatively close to the genes that they target, usually within 50,000 base pairs. It's relatively easy to visualise how this happens, through the long non-coding RNA and the Mediator complex acting to anchor the enzyme that copies DNA into messenger RNA. But there are a lot of situations where the enhancer and the protein-coding gene that it regulates are a really long distance apart on the chromosome, up to several million base pairs away. This is the difference between trying to pass the salt to someone who is on the other side of the table from you at breakfast, and trying to pass it to someone who is at the other end of a soccer field. It's quite difficult to visualise how this long-range interaction of gene and enhancer can happen. Neither the long non-coding RNA nor the Mediator complex is large enough to span such a huge distance.

In order to understand this process, we have to be a little more sophisticated than usual in the way we think of the genome. Much of the time it's very helpful to describe DNA in terms of a ladder, or railway lines, because it helps us to visualise the two strands and the way they are held together by base pairs. But the problem with this is it makes us think in very linear terms. We probably also think of DNA as being quite a stiff molecule because subconsciously we are comparing it with solid artefacts from our more familiar physical surroundings.

But we already recognise that DNA isn't a stiff molecule, because we know that it can be squashed up and compacted really dramatically to fit into the nucleus. So let's explore that a bit more. If we take the double-stranded nature of DNA as a given (so as not to complicate the picture) we can imagine a section of our genome as being like a very long piece of pasta, maybe the longest piece of tagliatelle ever created. This is marked in a couple of places by food dye, representing the enhancer and the protein-coding gene. Looking at Figure 12.3 we can see two scenarios. When the pasta is uncooked, it's very inflexible and the enhancer and gene are far apart. But if the pasta is cooked, the tagliatelle becomes flexible. It can fold and bend in all sorts of directions and these can bring the dyed regions representing the enhancer and gene together.

Some parts of our chromosomes are repressed and shut down almost permanently in different cells, to switch off genes that never need to be expressed in that tissue type. Our skin cells, for example, don't need to express the proteins that are used to carry oxygen around in the blood. These genomic regions are completely inaccessible in the skin cells, curled up tight like an over-wound spring. But there are huge regions that aren't in this hyper-condensed state and where genes are accessible and can potentially be switched on. In these sections the DNA is like the cooked pasta, like having the

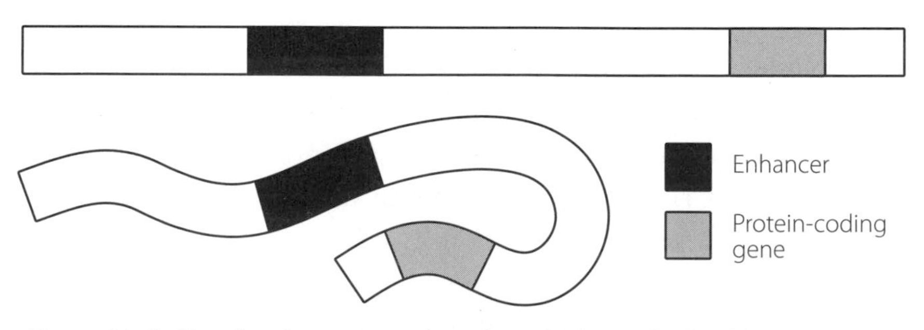

Figure 12.3 Simple schematic to show how folding of a flexible DNA molecule can bring two distant regions, such as an enhancer and a protein-coding gene, into close proximity to each other.

longest piece of tagliatelle in the world, filling an entire pot on its own. It bends and swirls in the cooking water, throwing out loops and arcs.

In this way, a protein-coding gene and its distant enhancer may come very close to each other. The long non-coding RNA and the Mediator complex then hold the two loops together and ensure that expression of the gene is driven up. Another complex also has to work with the Mediator complex to carry this out.* The additional complex is one that's also required for separating duplicated chromosomes during cell division, so it's well equipped for dealing with large-scale movements of DNA. Mutations in some of the genes that encode members of this additional complex cause two developmental disorders, called Roberts Syndrome and Cornelia de Lange syndrome.[17] The precise features of the affected children can be quite variable, probably depending on the exact gene that is mutated, and the precise mutation in that gene. Typically, the children are born small and remain relatively undersized; they have a learning disability and frequently present with limb abnormalities.[18]

The extent of this looping mechanism is quite remarkable, and may not just be restricted to enhancers. It may also be used to bring other regulatory elements close to genes. In a study of three human cell types, analysing just 1 per cent of the human genome, researchers identified over 1,000 of these long-range interactions in each cell line. The interactions were complex, most frequently involving regions that were separated by about 120,000 base pairs. Often the regulatory region looped up to a gene that wasn't the nearest one to it. In fact, in over 90 per cent of the loops, the nearest gene had been ignored. Think of this as needing to borrow a cup of sugar and visiting someone half a mile away instead of popping in to your next door neighbour.

* This additional complex is called Cohesin.

And if we continue the neighbour theme, the relationships were outrageously promiscuous. Imagine a 1970s partner-swapping party on steroids. Some genes interacted with up to twenty different regulatory regions. Some regulatory regions interacted with up to ten different genes. These probably don't all occur in the same cell at the same time. But what they show is that there is not a simple A to B relationship between genes and regulatory regions. Instead, there is a complex net of interactions, giving a cell or an organism an extraordinary amount of flexibility in how it regulates its overall tapestry of gene expression.[19] Although there is still plenty to be unravelled about the networks and how they operate, it would appear that while the junk DNA that forms promoters switches on our genomic engines, it's the junk DNA that forms long non-coding RNAs and enhancers that converts that engine from one powering a Sandero to one that can accelerate the Veyron down the freeway of life.

From cottage industry to the factory floor

Remarkable though the looping between individual regulatory regions and genes undoubtedly is, there is an even more dramatic set of long-range interactions that occur in cells. To understand the significance of this, a short social history lesson may be in order. In the early part of the 19th century in Britain, the bulk of all textile work was carried out as a cottage industry. Essentially, individuals worked in their homes on small-scale production. If you had mapped out the locations of textile production in a given region, you'd have a map with lots of individual dots on it, where each working cottage was located. Fast-forward about 50 years and into the Industrial Revolution and the same study would create a very different picture. Instead of a fairly homogenous dotted distribution, like a pointillist painting, you'd find a map with just a few large spots showing the location of big factories.

Even if we just think about the protein-coding genes, we know that thousands are typically switched on in a given human cell type. These genes are spread out across our 46 chromosomes, so we might expect that if we analysed a cell to visualise the geographical locations of the genes that are switched on we would see thousands of tiny dots spread throughout the nucleus. Instead, as shown schematically in Figure 12.4, there are about 300 to 500 larger spots.[20] Gene expression in our cells isn't a cottage industry. Instead it takes place in discrete locations in the nucleus known as factories.[21]

Each factory contains between four and 30 copies of the enzyme that makes a messenger RNA molecule from the DNA template, plus a large number of other molecules required to do the work.[22,23] The enzymes stay in one place and the relevant gene is reeled through to be copied.[24] In order for the gene to reach the factory, the DNA has to loop out to reach the right part of the cell nucleus. But the really ingenious bit is that more than one gene can be copied into messenger RNA at a time in the same factory. The combination of genes found in a single factory isn't random. The genes tend to be ones that code for proteins that are used for related functions in the cell. This is equivalent to having multiple

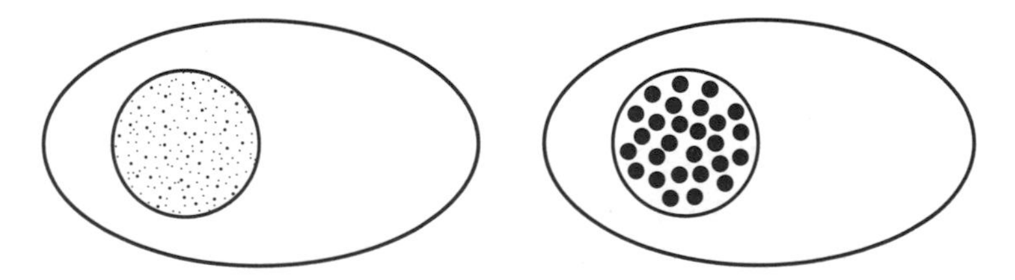

Figure 12.4 The dots represent the positions of protein-coding genes in the nucleus. If genes were positioned in the nucleus solely as a function of their position on chromosomes, we would see a diffuse pattern such as the one on the left. Instead, genes cluster together in three-dimensional space, creating a punctate pattern of gene localisation represented by the situation on the right.

parallel assembly lines in one physical factory. Once all the lines have completed their individual tasks, the factory can assemble the final product from the components. One factory produces boats, another builds food mixers. In our cells, the factories ensure that genes are expressed in a coordinated fashion. This means lots of loops unfurling from chromosomes and localising to the same regions of the nucleus simultaneously.

One example of this is a factory for the genes that code for the proteins required to create the complex haemoglobin molecule, which carries oxygen around in the blood.[25] Another factory is used to generate the proteins required in order to mount a strong immune response.[26] One important component of an effective immune response is the production of proteins called antibodies. Antibodies circulate in the blood and other body fluids, binding to any foreign matter that they detect. Scientists activated the cells that produce antibodies and then studied how certain key genes looped out. The genes they analysed were the ones required to create antibody molecules. They found that these key genes moved to the same factory as each other. Remarkably, some of these genes were completely physically separate from each other normally, as they are located on different chromosomes.

Although this is a remarkable way of coordinating gene expression, it may also carry risks. Burkitt's lymphoma is the aggressive cancer we met earlier in this chapter. The cell type that becomes abnormal in this disease is the cell type that produces antibodies. In this condition, a strong promoter from one chromosome gets abnormally positioned next to a gene from another chromosome. Until recently we didn't understand why these regions were susceptible to joining up, because we thought of them as being physically distant from each other, as they are on different chromosomes. But now we know that the regions that 'swap' to create the dangerous abnormal hybrid chromosome are both regions that move to the factory described in the previous paragraph.

This might be how the two different chromosomes get close enough together to swap their material, perhaps if both break simultaneously and are wrongly repaired when in the factory.

While it might seem that evolution would have selected against this dangerous situation, we need to remember yet again that natural selection is about compromise, not perfection. The advantages of producing antibodies to fight off infections and thereby keep us alive long enough to reproduce clearly outweigh the potential disadvantages of an increased cancer risk.

13. No Man's Land

When we think about the First World War, the prevailing image many of us have is probably of men in trenches. Opposing armies dug into the muddy landscapes for the ultimate exposition of war as months of boredom, punctuated by moments of acute terror.[1] The trenches occupied by the armies were separated by stretches of terrain that were not under the control of either combatant. These stretches were 'No Man's Land', and could be as narrow as a couple of hundred metres, or over a kilometre wide. At night, the soldiers would creep out of the trenches for reconnaissance, to lay barbed wire and to retrieve injured or dead compatriots.

The human genome contains multiple regions of No Man's Land, keeping different elements apart from each other. Just like the quagmires of the First World War, these genomic barriers vary in size and are fairly fluid, depending on where they lie in relation to their troop movements. And just like the No Man's Land of Europe in those awful few years of slaughter, these regions are anything but devoid of activity. The No Man's Land of the human genome binds proteins, garners epigenetic modifications and regulates the interactions of different genetic elements in a highly active way.

This is important to our cells, because most of our genes are all over the place.*[2] By this we mean that genes are scattered around on our 23 pairs of chromosomes in a fairly nonsensical way. As we have already seen, the genes that code for the proteins required to

---

* There are a couple of exceptions to this where genes are clustered in a way that reflects their expression patterns. The main ones are the HOX genes that control body patterning and the Ig genes that code for antibodies.

make haemoglobin are brought together by changes in the three-dimensional arrangements of chromosomes. This compensates for the fact that they aren't arranged next to each other in a nice, neat way. If we look at how most of our genes are distributed, they are like the donations to a jumble sale or charity shop before they've been organised sensibly.

This can mean that our cells contain a gene that codes for a protein required in the foetal liver next to a gene for a protein expressed in the adult skin. There's a huge number of such situations and this creates potential difficulties. It means that our cells require barriers between different elements to maintain different patterns of gene expression. The control needs to be relevant to a specific cell type, and to the particular developmental stage. We don't want teeth genes expressed in our eyes or heart genes expressed in our bladders.

We know that epigenetic modifications influence gene expression. If we take the brain as an example, there are some genes that are never expressed in neuronal cells. For instance, the protein keratin is used in hair and nails, but isn't used by our adult grey matter. In brain cells, the keratin gene is switched off and it's kept in an inactive state by a particular pattern of epigenetic modifications. But as we've already seen, epigenetic modifications are blind to DNA sequence. What's to stop these repressive modifications from creeping along from the keratin gene and starting to switch off other genes as well?

This is particularly a problem because epigenetic modifications are often self-sustaining. Let's take the case of modifications that are involved in repressing gene expression. These modifications attract other proteins that reinforce the initial change, making it even harder to reactivate gene expression. These in turn can attract proteins that continue to add repressing epigenetic modifications, to prevent escape from inactivation. But we can imagine that the borders of the repression are quite vague, because the epigenetic machinery doesn't recognise specific DNA sequences. So, at the periphery of the repressed regions, the epigenetic modifications could spread out.

Halting the spread

Our cells have evolved a remarkable way to prevent this. Just as fire crews will cut down stands of trees or blow up buildings to create a gap in the path of an inferno, our genome removes the fuel for the epigenetic machinery. Junk DNA that acts as an insulator between repressed and active regions of the genome loses its histone proteins. No histone proteins means no epigenetic histone modifications. No modifications means no spreading of epigenetic activity. This stops repressive modifications creeping into active genes and also prevents the opposite effect. This is shown in Figure 13.1.

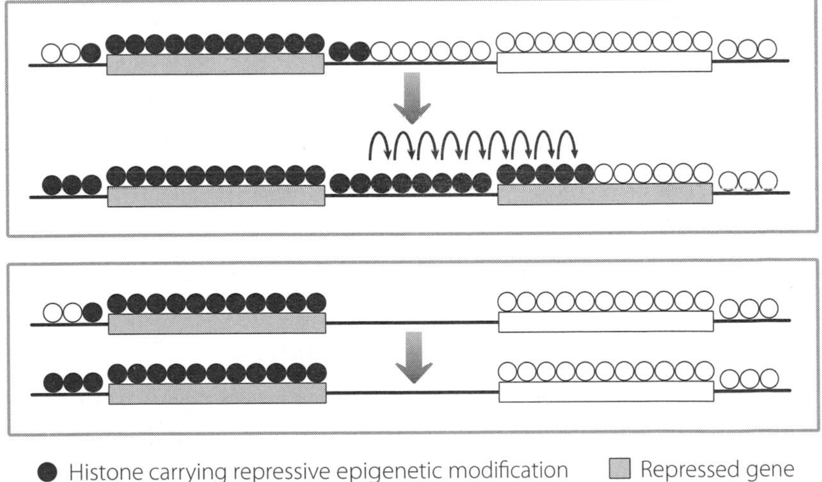

● Histone carrying repressive epigenetic modification
○ Histone without repressive modification
■ Repressed gene
□ Expressed gene

Figure 13.1 In the upper panel, repressive modification patterns spread from one gene to the next. In the lower panel, the lack of histones in the insulator regions between two genes prevents the spread of the repressive epigenetic modifications, and stops the right-hand gene from being abnormally silenced.

But because different cells need to insulate different regions (we do, after all, want keratin expressed in the cells that create hair) we can deduce that DNA sequence alone isn't enough to create an insulator. Instead, these are generated by complex, situationally dependent interactions between the genome and the combinations of proteins expressed in a cell at any one time.

One of the most important of these proteins is a ubiquitously expressed one that we can refer to as 11-FINGERS.* It's a large, highly conserved protein with a characteristic structure. The way that it folds in three dimensions means that there are eleven finger-like projections that stick out from the protein. Each of these eleven fingers can recognise a defined DNA sequence, but not each finger recognises the same sequence.

Imagine an eleven-fingered pianist wearing gloves where the wool on each digit is one of four colours. Combine this with a piano where each key is also one of the same four colours, assigned randomly between the keys. The rules are that the pianist can play any notes she likes, but must always hit between two and eleven notes simultaneously, and the colours on the fingers and keys must match. We can start to see that there are an awful lot of possible combinations. And to understand the extent of the different options, now imagine that the piano has thousands of keys.

The 11-FINGER protein is able to bind to lots of different genomic sequences in a similar way. It can bind to tens of thousands of sites in human cells. In addition to binding itself to DNA, 11-FINGER also binds other proteins. We can again invoke our abnormally digited piano player to visualise this. Imagine there is Velcro on the backs of the gloves, which can bind fuzzy balls of fluff. The coloured fingers of the gloves hit the piano keys, the backs of the gloves get covered in fluffy fabric balls.

So it is for 11-FINGER. The finger-like projections bind to

* The formal name for this is CTCF.

DNA, the other surfaces of the protein bind other proteins. The precise binding partners will depend on the complement of proteins being expressed in a cell. One of the proteins can alter the coiling of DNA, which can be important for controlling gene expression.[3] Another is a protein that deposits specific epigenetic modifications.[4] In some regions the types of genomic interlopers we met in Chapter 4 serve as insulators, preventing the spread of activating or repressive epigenetic modifications from one region to another.[5]

Some tRNA genes can act as insulators. They can stop expression of one gene driving inappropriate expression of a neighbouring gene. This is an additional benefit of having lots of tRNA genes, which demonstrates the economical way with which evolution has made the most of raw material.

The way this works is shown in Figure 13.2. A classical protein-coding gene is coated with epigenetic modifications that promote its expression. The enzyme that binds to this gene and copies it into RNA (which will ultimately be processed to form mature messenger RNA) can be a bit of a runaway train: once it starts copying it tends to keep going. If there is another protein-coding gene nearby, the enzyme could keep going and copy this as well. But if there are two or more tRNA genes in between, this won't happen. tRNA genes are switched on pretty much all the time, because they are involved in the creation of all proteins. There is an enzyme that copies tRNA genes to create tRNA molecules from the DNA template. But this is different from the enzyme that carries out a similar job to generate messenger RNA molecules from classical protein-coding genes. The enzyme that creates the tRNA molecules acts like a big burly bouncer, stopping the other enzyme from getting through the door to the next gene. Because the enzyme that copies tRNA genes can't bind to classical protein-coding genes, this keeps the overall gene expression in this region under tight spatial control.[6]

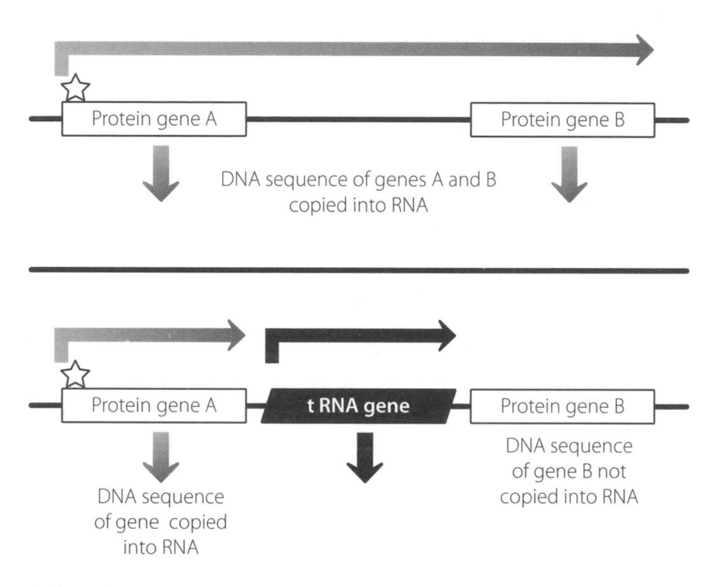

Figure 13.2 The enzyme that copies DNA into messenger RNA from protein-coding genes binds at the star at the start of gene A. If nothing stops it, the enzyme could keep on copying until it has also copied protein-coding gene B into messenger RNA, perhaps inappropriately. tRNA genes are copied from DNA into functional tRNA molecules by a different enzyme. This blocks the progress of the enzyme creating messenger RNA from gene A, and prevents inappropriate use of gene B.

Because there has been such an emphasis in biology on the dividends from the development of DNA sequencing technologies, it's always tempting to think that most of the big conceptual breakthroughs arise from high-end molecular approaches. But the reality is that basic human biology and logical thought actually take us a long way.

Why XX is different from XXX

In Chapter 7 we saw that female mammals always inactivate one X chromosome in their cells, to ensure that they have the same levels of X chromosome gene expression as male cells. Our cells are able to count. If a female cell contains three X chromosomes, it will switch off two of them. Conversely, if there is only one X chromosome, the cell leaves this switched on.

This leads us to a pretty obvious prediction. It doesn't matter how many X chromosomes a cell contains, because X inactivation will always ensure that only one is functionally active. Therefore, as long as a person contains at least one X chromosome in each cell, they will be completely normal and healthy.

The problem is, this isn't true. Women with only one X chromosome, or with three X chromosomes, do have detectable symptoms. So do men who have two X chromosomes in addition to their Y. One explanation could be that maybe X inactivation isn't working well in these people, but that doesn't seem to be the case. X inactivation is a very robust system. It's unlikely to work perfectly every single time – nothing else in biology does. But random inadequacies in the system wouldn't explain why all women with just one X chromosome present with very similar clinical symptoms.

Women with just one X chromosome are shorter than average, and have underdeveloped ovaries.[7] Women with three X chromosomes are taller than average and at increased risk of learning disabilities and developmental delay as children.[8] Males with two X chromosomes (plus a Y of course) are taller than average, and may have relatively small testicles, leading to problems caused by low production of the male hormone, testosterone. They are also at increased risk of learning disabilities.[9]

Although potentially distressing for the patients and their families, the symptoms are milder than we see for patients with abnormal numbers of autosomal chromosomes (remember Down's, Edward's and Patau sydromes – *see pages* 76–7). That's because although the X chromosome is large, most of the genes on it are appropriately inactivated, no matter how many copies of this chromosome are present. But there are some that aren't.

To understand what is happening, we need to think back to what happens when eggs or sperm are created. At a certain stage, the chromosomes line up in pairs and then one of each pair is pulled to opposite ends of the cell. The cell divides and its daughter cells contain one of each pair. In a female cell this is easy to visualise. The two X chromosomes pair up and then can be separated, in exactly the same way as any other pair of chromosomes from number 1 to number 22. But when males are creating sperm, there is a problem. Males contain one large X chromosome and one tiny Y chromosome. These are very different from each other. Yet somehow, during the creation of sperm, the X and the Y must find each other and pair up, despite being so different.

The reason they can do this is because there is a small region at the ends of the X and Y chromosomes where they are very similar to each other. This allows them to recognise each other and to associate during cell division, holding hands until they need to move to opposite ends of the dance floor.

These stretches are known as pseudoautosomal regions. They contain protein-coding genes, and they are protected from silencing during X inactivation. The genes in the pseudoautosomal region are treated very differently from most of the other genes on the X chromosome. This pattern of activated and inactivated genes, which leads to detectable symptoms in males and females with the wrong number of X chromosomes, was a clear sign from biology that cells contain very fundamental ways of functionally separating different blocks of DNA.

X inactivation is critically dependent on the Xist long non-coding RNA spreading along the chromosome on which it is expressed. But Xist doesn't spread into the pseudoautosomal regions. The protection from this in the pseudoautosomal region shows us that our genomes have evolved in such a way that at key positions, they can draw a line in the sand. As Jean-Luc Picard declared, in reference to Borg incursions into Federation space, 'The line must be drawn here! This far, no farther!'[10] Junk insulator regions prevent the creeping genomic paralysis that spreads out from the Xist locus.

Figure 13.3 shows how these non-silenced regions result in changes in people who have the wrong numbers of X chromosomes. A woman who only has one X chromosome expresses 50 per

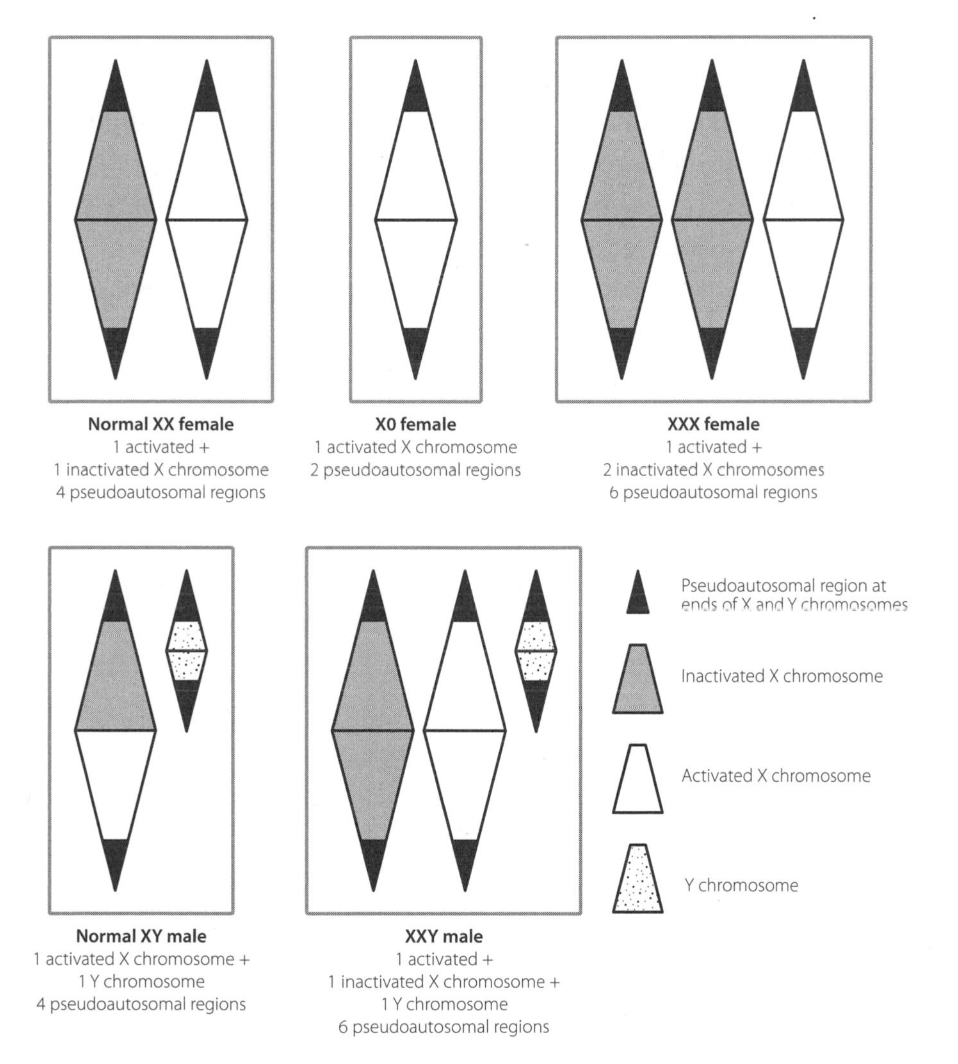

Figure 13.3 The effects of different numbers of X chromosomes in male and female cells. Because of X inactivation, there is only one active X chromosome in each cell. But because the pseudoautosomal regions at the ends of the X and Y chromosomes escape X inactivation, their numbers increase or decrease pathologically with changes in X chromosome number.

cent of the normal amounts of gene products from the pseudo-autosomal regions as a typical XX woman. A woman with three X chromosomes produces 50 per cent more of these gene products than normal, as does a male with two X chromosomes and a Y.

It's no coincidence that both males and females with an extra X chromosome are taller than average, and women lacking an X tend to be on the short side. The pseudoautosomal region contains a particular protein-coding gene*[11] which controls the expression of other genes and is important for development of the skeleton, especially the long bones of the arms and legs. Men and women with extra X chromosomes express more of this protein than normal, which tends to increase leg length and hence height. The opposite is true for women lacking an X chromosome. It's one of the few examples in the human genome where we can really identify a single region which has a significant impact on the normal range of human height. Outside of this region, height is influenced by multiple sites in the genome,[12] and many of these are regions of junk DNA, where we don't yet know how they individually contribute to making you a Harlem Globetrotter, or someone who is always overlooked in a bar.

* This protein is known as SHOX or 'short stature homeobox'.

14. Project ENCODE – Big Science Comes to Junk DNA

If you ever find yourself far from city lights, on a cloudless night with no moon, grab a blanket and lie on the ground and look up at the stars. It's one of the most wonderful sights imaginable, and quite breathtaking for anyone who spends their life in a city. The glints of silver in the dark blanket of the heavens seem too many to count.

But – if you have access to a telescope, you realise that there is so much more in the firmament than you can detect with the naked eye. There are details like the rings of Saturn, and there are vastly more stars than we could ever imagine. There is so much more in the apparent darkness of the universe than can be seen just with our limited unaided vision. This becomes even more obvious if we use equipment that can detect the energy in the other parts of the electromagnetic spectrum, beyond just the visible wavelengths. More information keeps pouring in, from gamma waves to the microwave background. Those details and those stars have always been there, we just couldn't detect them when we relied only on eyesight.

In 2012, a whole slew of papers was published that attempted to turn a telescope onto the furthest reaches of the human genome. This was the work of the ENCODE consortium, a collaborative effort involving hundreds of scientists from multiple different institutions. ENCODE is an acronym derived from *Enc*yclopaedia *O*f

*D*NA *E*lements.[1] Using the most sensitive techniques available, the researchers probed multiple features of the human genome, analysing nearly 150 different cell types. They integrated the data in a consistent way, so that they could compare the outputs from the different techniques. This was important because it's very difficult to make comparisons between data sets that have been generated and analysed differently from each other. Such piecemeal data were what we had previously relied on.

When the ENCODE data were published, there was an enormous amount of attention from the media, and from other researchers. Press coverage included headlines such as 'Breakthrough study overturns theory of "junk DNA" in genome';[2] 'DNA project interprets "Book of Life"'[3] and 'Worldwide army of scientists cracks the "junk DNA" code'.[4] We might imagine that other scientists would all be congratulatory, and even grateful for all the additional data. And a lot were really fascinated, and are using the data every day in their labs. But the acclaim has been far from universal. Criticism has come mainly from two camps. The first is the junk sceptics. The second is the evolutionary theorists.

To understand why the first group were upset, we can examine one of the pithiest statements in the ENCODE papers:

> These data enabled us to assign biochemical functions for 80% of the genome, in particular outside of the well-studied protein-coding regions.[5]

In other words, instead of being mainly dark sky with less than 2 per cent of the space occupied by stars, ENCODE was claiming that in our genome four-fifths of the celestial canopy is filled with objects. Most of the objects aren't stars, assuming stars represent the protein-coding genes. Instead, they could be asteroids, planets, meteors, moons, comets and any other interstellar objects you can think of.

As we have seen, many research groups had already assigned functions to some of the dark area, including promoters, enhancers, telomeres, centromeres and long non-coding RNAs. So most scientists were comfortable with the idea that there was more to our genome than the small proportion that encoded proteins. But 80 per cent of the genome having a function? That was a really bold claim.

Although startling, these data had been foreshadowed by indirect analyses in the previous decade by scientists trying to understand why humans are so complicated. This was the problem by which so many people had been puzzled ever since the completion of the human genome sequence failed to find a larger number of protein-coding genes in humans than in much simpler organisms. Researchers analysed the size of the protein-coding part of the genome in different members of the animal kingdom and also the percentage of the overall genome that was junk. The results, which we touched on in Chapter 3, are shown in Figure 14.1.

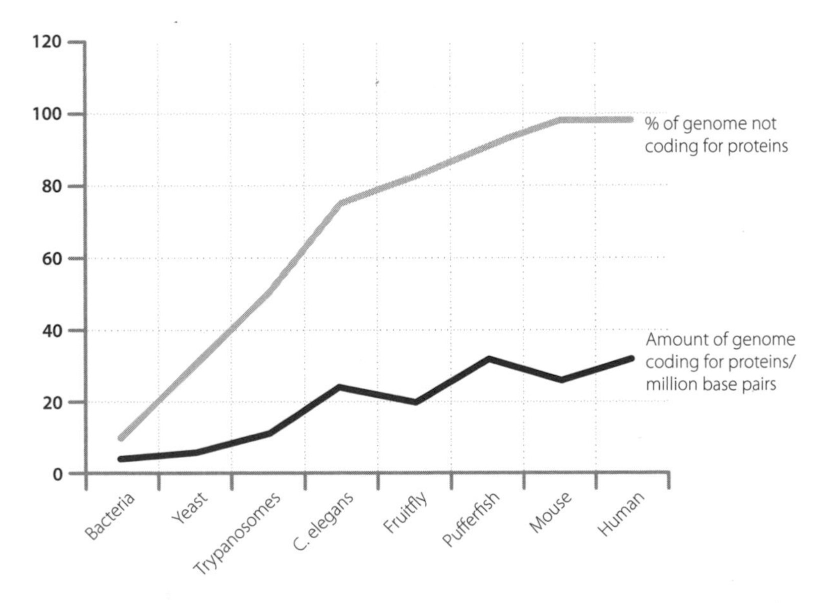

Figure 14.1 Graphical representation showing that organismal complexity scales more with the proportion of junk DNA in a genome than with the size of the protein-coding part of the same genome.

As we have seen, the amount of genetic material that codes for proteins doesn't scale very well with complexity. There is a much more convincing relationship between the percentage of junk in the genome and how complicated an organism is. This was interpreted by the researchers as suggesting that the difference between simple and complex creatures is mainly driven by junk DNA. This in turn would have to imply that a significant fraction of the junk DNA has function.[6]

Multiple parameters

ENCODE calculated its figures for level of function in our genome by combining all sorts of data. These included information on the RNA molecules that they detected. These were both protein-coding and ones that didn't code for protein, i.e. junk RNAs. They ranged in size from thousands of bases to molecules a hundred times smaller. ENCODE also defined genome regions as being functional if they carried particular combinations of epigenetic modifications that are usually associated with functional regions. Other methodologies involved analysing regions that looped together in the way that we encountered in the previous chapter. Yet another technique was to characterise the genome in terms of specific physical features associated with function.*

These features varied across the different human cell types analysed, reinforcing the concept that there is a great deal of plasticity in how cells can use the same genomic information. For example, analyses of looping found that any one specific interaction between different regions was only detected in one out of three cell types.[7] This suggests that the complex three-dimensional folding of our genetic material is a sophisticated, cell-specific phenomenon.

* These were typically accessibility to enzymes that can cut DNA molecules, which is a sign of an open structure that may be able to be copied into RNA.

When looking at the physical characteristics that are typically associated with regulatory regions, researchers concluded that these regulatory DNA regions are also activated in a cell-dependent manner, and in turn that this junk DNA shapes cell identity.[8] This conclusion was reached after the scientists identified nearly 3 million such sites from analysis of 125 different cell types. This doesn't mean that there were 3 million sites in each cell type. It means that 3 million were detected when the different sites from each cell type were added up. Yet again, this suggests that the regulatory potential of the genome can be used in different ways, depending on the needs of a specific cell. The distribution of the sites among different cell types is shown in Figure 14.2.

Over 90 per cent of the regulatory regions identified by this method were more than 2,500 base pairs away from the start of the nearest gene. Sometimes they were far from any gene at all, in other cases they were in a junk region within a gene body, but still far from the beginning.

Figure 14.2 Researchers analysing the ENCODE data sets identified over 3 million sites with the characteristics of regulatory regions, when they assessed multiple human cell lines. The areas of the circles in this diagram represent the distribution of these sites. The majority were found in two or more cell types, although a large fraction was also specific to individual cell types. Only a very small percentage were found in every cell line that was analysed.

Most gene promoters were associated with more than one of these regions, and each region was typically associated with more than one promoter. Yet again, it appears that our cells don't use straight lines to control gene expression, they use complex networks of interacting nodes.

Some of the most striking data suggested that over 75 per cent of the genome was copied into RNA at some point in some cells.[9] This was quite remarkable. No one had ever anticipated that nearly three-quarters of the junk DNA in our cells would actually be used to make RNA. When they compared protein-coding messenger RNAs with long non-coding RNAs the researchers found a major difference in the patterns of expression. In the fifteen cell lines they studied, protein-coding messenger RNAs were much more likely to be expressed in all cell lines than the long non-coding RNAs, as shown in Figure 14.3. The conclusion they reached from this finding was that long non-coding RNAs are critically important in regulating cell fate.

Taken in their entirety, data in the various papers from the

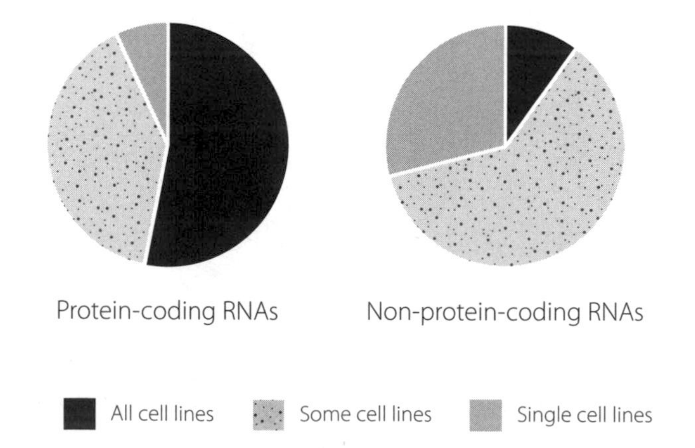

Figure 14.3 Expression of protein-coding and non-coding genes was analysed in fifteen different cell types. Protein-coding genes were much more likely to be expressed in all cell types than was the case for regions that produced non-coding RNA molecules.

ENCODE consortium painted a picture of a very active human genome, with extraordinarily complex patterns of cross-talk and interactivity. Essentially the junk DNA is crammed full of information and instructions. It's worth repeating the hypothetical stage directions from the Introduction: 'If performing *Hamlet* in Vancouver and *The Tempest* in Perth, then put the stress on the fourth syllable of this line of *Macbeth*. Unless there's an amateur production of *Richard III* in Mombasa and it's raining in Quito.'[10]

This all sounds very exciting, so why was there a considerable degree of scepticism about how significant these data are? Part of the reason is that the ENCODE papers made such large claims about the genome, particularly the statement that 80 per cent of the human genome is functional. The problem is that some of these claims are based on indirect measures of function. This was especially true for the studies where function was inferred either from the presence of epigenetic modifications or from other physical characteristics of the DNA and its associated proteins.

Potential versus actual

The sceptics argue that at best these data indicate a *potential* for a region to be functional, and that this is too vague to be useful. An analogy might help here. Imagine an enormous mansion, but one where the owners have fallen on hard times and the power has been disconnected. Think Downton Abbey in the hands of a very bad gambler. There could be 200 rooms and five light switches in each room. Each switch could potentially turn on a bulb, but it may be that some of the switches were never wired up (aristocrats are not known for their electrical talents), or the associated bulb is broken. Just because the switches are on the walls, and can be flicked between the on and off positions, it doesn't actually tell us that they will really make a difference to the level of light in the room.

The same situation may take place in our genomes. There may be regions that carry epigenetic modifications, or have specific physical characteristics. But this isn't enough to demonstrate that they are functional. These characteristics may simply have developed as a side effect of something that happened nearby.

Look at any photo of Jackson Pollock creating one of his abstract expressionist masterpieces.[11] It's a pretty safe bet that the floor of his studio got spattered with paint as he created his canvases. But that doesn't mean that the paint spatters on the floor were part of the painting, or that the artist endowed them with any meaning. They were just an inevitable and unimportant by-product of the main event. The same may be true of physical changes to our DNA.

Another reason why some observers have been sceptical about the claims made in ENCODE is based on the sensitivity of the techniques used. The researchers were able to use methods that were far more sensitive than those employed when we first started exploring the genome. This allowed them to detect very small amounts of RNA. Critics are afraid that the techniques are too sensitive and that we are detecting background noise from the genome. If you are old enough to remember audio tapes, think back to what happened if you turned the volume on your tape deck really high. You usually heard a hissing noise behind the music. But this wasn't a sound that had been laid down as part of the music, it was just an inevitable by-product of the technical limitations of the recording medium. Critics of ENCODE believe that a similar phenomenon may also occur in cells, with a small degree of leaky expression of random RNA molecules from active regions of the genome. In this model, the cell isn't actively switching on these RNAs, they are just being copied accidentally and at very low levels because there is a lot of copying happening in the neighbourhood. A rising tide lifts all boats, but it will also raise any old bits of wood and abandoned plastic bottles that happen to be in the water as well.

This seems like a quite significant problem when we realise that in some cases the researchers detected less than one molecule of a particular RNA per cell. It isn't possible for a cell to express somewhere between zero copies and one copy of an RNA molecule. A single cell either produces no copies of that particular RNA or it makes one or more copies. Anything else is like being 'sort of' pregnant. You're either pregnant or you're not, there's nothing in between.

But this doesn't actually tell us that the techniques used were too sensitive. Instead, it tells us that our techniques still aren't sensitive enough. Our methods aren't good enough to allow us to isolate single cells and analyse all the RNA molecules in that cell. Instead we have to rely on isolating multiple cells, analysing all the RNA molecules and then calculating how many molecules were present on average in the cells.

The problem with this is that it means we can't distinguish between a large percentage of the cells in a sample each expressing a small amount of a specific RNA and a small percentage of the cells each expressing a large amount of the RNA. These different scenarios are shown in Figure 14.4.

The other problem we have is that we have to kill the cells in order to analyse the RNA molecules. As a consequence we only generate snapshots of the RNA expression where ideally we would want to have the equivalent of a movie so that we could see what was happening to RNA expression in real time. The problem inherent in this is demonstrated in Figure 14.5.

Ideally, of course, we should be able to test if the findings of ENCODE really stand up to scrutiny by direct experiments. But this takes us back to the problem that there were so many findings. How do we decide on candidate regions or RNA molecules to study? The additional complication is that many of the features identified in the ENCODE papers are parts of large, complex networks of interactions. Each component may have a limited effect on its own in the overall picture. After all, if you cut through one

2	2	2	2	2	2
2	2	2	2	2	2
2	2	2	2	2	2
2	2	2	2	2	2
2	2	2	2	2	2
2	2	2	2	2	2

Number of RNA molecules detected = 72

36	0	0	0	0	0
0	0	0	0	0	0
0	0	0	0	0	0
0	0	36	0	0	0
0	0	0	0	0	0
0	0	0	0	0	0

Number of RNA molecules detected = 72

Figure 14.4 Each small square represents an individual cell. The figures inside the cell indicate the number of molecules of a specific RNA molecule produced in that cell. The researcher analyses a batch of cells, due to the sensitivity limits of the detection methods available. This means that the researcher only has access to the total number of molecules in the batch and cannot distinguish between (*left*) 36 cells all containing two molecules and (*right*) two cells (out of 36) each containing 36 molecules – or any other combination that results in an overall total of 72 molecules.

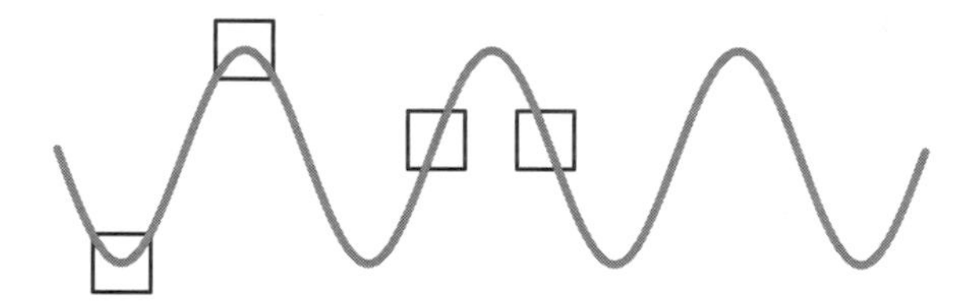

Figure 14.5 Expression of a specific RNA in a cell may follow a cyclical pattern. The boxes represent the points at which a researcher samples the cells to measure expression of that RNA molecule. The results may appear very different when comparing different batches of cells, perhaps from discrete tissues, but it may be that this simply reflects a temporal fluctuation rather than a genuine, biologically significant variation.

knot in a fishing net, you won't destroy the overall function. The hole may allow the occasional fish to escape but losing one small fish probably won't have too much impact on your catch. Yet that doesn't mean all the knots are unimportant. They all are important, because they work together.

The evolutionary battleground

The authors of the ENCODE papers, and of the accompanying commentaries, also used the data to draw evolutionary conclusions about the human genome. Part of the reason for this lay in an apparent mismatch. If 80 per cent of the human genome has function, you would predict that there should be a significant degree of similarity between the human genome and at the very least the genomes of other mammals. The problem is that only about 5 per cent of the human genome is conserved across the mammalian class, and the conserved regions are highly biased towards the protein-coding entities.[12] In order to address this apparent inconsistency, the authors speculated that the regulatory regions have evolved very recently, and mainly in primates. Using data from a large-scale study of DNA sequence variation in different human populations, they concluded that the regulatory regions have relatively low diversity in humans, whereas the diversity is much higher in areas that have no activity at all. One of the commentaries explored this further, using the following argument. Protein-coding sequences are highly conserved in evolution because a particular protein is often used in more than one tissue or cell type. If the protein changed in sequence, the altered protein might function better in a particular tissue. But that same change might have a really damaging effect in another tissue that relies on the same protein. This acts as an evolutionary pressure that maintains protein sequence.

But regulatory RNAs, which don't code for proteins, tend to be more tissue-specific. Therefore they are under less evolutionary pressure because only one tissue relies on a regulatory RNA, and possibly only during certain periods of life or in response to certain environmental changes. This has removed the evolutionary brakes on the regulatory RNAs and allowed us to diverge from our mammalian cousins in these regions. But across human populations, there has been pressure from evolution to maintain the optimal sequence for these regulatory RNAs.[13]

Biologists tend to be a rather restrained social group when it comes to disagreements. There's the occasional aggressive question-and-answer session at a conference but generally public pronouncements are carefully phrased. This is especially true of anything that is published, rather than said at a meeting. We all know how to read between the lines, of course, as shown in Figure 14.6 but typically, published papers are carefully phrased. That's what made the debate that followed ENCODE particularly entertaining to the relatively disinterested observer.

The most forthright responses were mainly from evolutionary biologists. This wasn't altogether surprising. Evolution is the biological discipline where emotions tend to run highest. Normally the bullets are targeted at creationists, but the Gatling guns may also be turned on other scientists. Epigeneticists working on the transmission of acquired characteristics from parent to offspring were probably quite relieved that ENCODE took them out of the firing line for a while.[14]

The angriest critique of ENCODE included the expressions 'logical fallacy', 'absurd conclusion', 'playing fast and loose' and

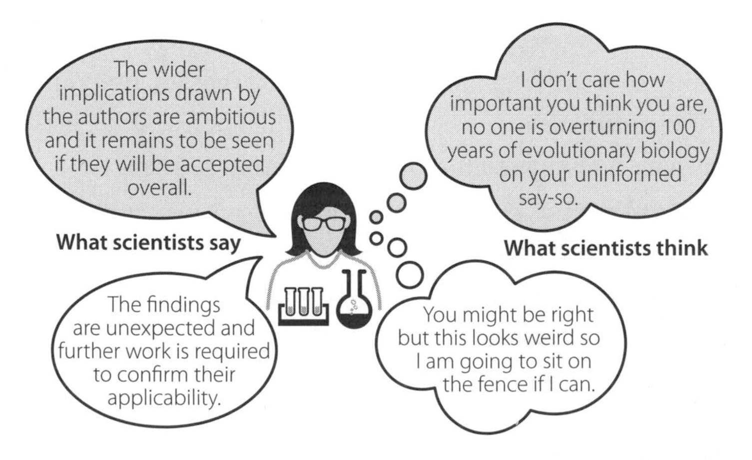

Figure 14.6 Scientists are usually outwardly polite (left-hand statements), but are sometimes just speaking in barely disguised code (right-hand thoughts) …

'used the wrong definition wrongly'. Just in case we were still in doubt about their direction of travel, the authors concluded their paper with the following damning blast:

> The ENCODE results were predicted by one of its lead authors to necessitate the rewriting of textbooks. We agree, many textbooks dealing with marketing, mass media hype, and public relations may well have to be rewritten.[15]

The main criticisms from this counter-blast centred around the definition of function, the way that the ENCODE authors analysed their data, and the conclusions drawn about evolutionary pressures. The first of these applied to the problems we have already described, using our Jackson Pollock and Downton Abbey analogies. In some ways, these problems derive in large part from difficulties in separating mathematics from biology. The ENCODE data sets were predominantly interpreted by the original authors through the use of statistical and mathematical approaches. The sceptics argue that this leads us down a blind alley, because it doesn't take into account biological relationships, and that these are critically important. They use a very helpful analogy to explain this. The reason the heart is important is that it pumps blood around the body. That's the biologically important relationship. But if we analysed the actions of the heart just by a mathematically derived map of its interactions, we would draw some ridiculous conclusions. These could include that the heart is present so that it can add weight to the body, and to produce the sound 'lub-dub'. These are both things that the heart undoubtedly does, but they are not its function. They are just contingent on its genuine role.

The authors criticised the analytical methods because they felt that the ENCODE teams had not been consistent in the way they applied their algorithms. One consequence of this was that effects seen in a large region might weigh down an analysis inappropriately. For example, if a block of 600 base pairs was classified as being functional, when all the work was actually carried out by just ten of them, this would dramatically skew the percentage of the genome that would be designated as having a function.

The evolutionary argument was that the ENCODE authors ignored the standard model that regions with large amounts of variation are reflective of a lack of evolutionary selection, which in turn means they are relatively unimportant. If you want to overturn such a long-held principle, you need to have very strong grounds for doing so. But the critics claimed that the ENCODE papers, although containing huge amounts of data, had only focused on an inappropriately small number of regions when drawing evolutionary conclusions from the sequences of humans and other primates.

There are interesting scientific arguments on both sides, but it would be disingenuous to believe that the amount of heat and emotion generated by ENCODE has been purely about the science. We can't ignore other, very human factors. ENCODE was an example of Big Science. These are typically huge collaborations costing millions and millions of dollars. The science budget is not infinite and when funds are used for these Big Science initiatives, there is less money to go around for the smaller, more hypothesis-driven research.

Funding agencies work hard to get the balance right between the two types of research. In many cases, Big Science is funded if it generates a resource that will stimulate a great deal of other science. The original sequencing of the human genome would be a clear example of this, although we should recognise that even that was not without its critics. But with ENCODE the controversy is not around the raw data that were generated, it's about how those data are interpreted. That makes it different from a pure infrastructure investment in the eyes of the critics.

When all stages and aspects of ENCODE are added up, it cost in the region of a quarter of a billion dollars. The same amount of money could have funded at least 600 average-sized single research grants focusing on investigation of individual hypotheses. Choosing how to distribute funding is a balancing act, and at these levels of funding it is guaranteed to create division and concern.

A company called Gartner created a graphic that shows how new technologies are perceived. It is known as the Hype Cycle. At first everyone is very excited – 'the peak of inflated expectations'. When the new tech fails to transform everything about your life there is a crash leading to the 'trough of disillusionment'. Eventually, everyone settles down, there is a steady growth in rational understanding and finally a productive plateau is reached.

With something like ENCODE this cycle is extraordinarily compressed, because of the polarisation from the most vocal groups. Those scientists with inflated expectations are operating at exactly the same time as those in the trough. Pretty much everyone else is pragmatic, and will use the data from ENCODE when it is useful to do so. Which is usually when it can help inform a specific question that an individual scientist finds interesting.

15. Headless Queens, Strange Cats and Portly Mice

The ENCODE consortium identified a daunting abundance of potentially functional elements in the human genome. Given the huge numbers, it's hard to define a sensible strategy for deciding which candidate regions to experiment on first. But the task may not be quite as difficult as it seems, and that's because, as always, nature has decided to point the way. In recent years scientists have begun to identify human diseases that are caused by tiny changes to regulatory regions of the genome. Previously, these might have been dismissed as harmless random variations in junk DNA. But we now know that in some cases just a single base-pair change in an apparently irrelevant region of the genome can have a definite effect on an individual. In rare cases, the effect is so severe that life itself is impossible.

We'll start with a less dramatic example, but one that takes us back about 500 years, to the reign of King Henry VIII in England. Most British schoolchildren are at some point taught a useful rhyme to help them remember what happened to the six wives of this notorious monarch:

Divorced, beheaded, died,
Divorced, beheaded, survived.

(Feel free to send a thank-you email when this handy little ditty helps you in a quiz.)

The first wife to be beheaded was Anne Boleyn, the mother of the future Queen Elizabeth I. After her death, the Tudor spin doctors launched quite a smear campaign and Anne Boleyn's physical appearance was described in such a way that she sounded like the 16th-century image of a witch. She was characterised as having a projecting tooth, a large mole under her chin and six fingers on her right hand. The story of that extra finger has passed down in folklore, although there is little if any evidence that it was true.[1]

Perhaps one of the reasons that the story has been accepted is because it's not completely ludicrous. It's not as if the chroniclers claimed that the former queen had three legs. There are people who are born with an extra finger, although usually they have an extra finger on each hand rather than just one.

There is a protein-coding gene that is very important in the correct development of the hands and feet.* The protein acts as a morphogen, meaning that it governs patterns of tissue development. The effects of the protein are very dependent on its concentration, and in the developing embryo there is a gradient effect, where high levels in one region gradually fade away to lower levels in adjacent tissues.

Mittens and kittens

One of the features controlled by this morphogen is the number of fingers. If the expression levels of the protein are wrong, babies are born with extra fingers. Over ten years ago researchers discovered that some cases of extra fingers were caused by a tiny genetic change. This wasn't in the morphogen gene, but in a region of junk DNA about a million base pairs away. They

* The protein is called Sonic Hedgehog, symbol SHH. Researchers went through a phase of giving genes apparently comic names. This is now discouraged as it's suddenly not so amusing if a genetic counsellor has to pass on a whimsical gene name to the parents of a child with a severe genetic condition.

identified the change in a huge Dutch family where the presence of extra fingers was clearly inherited as a genetic trait. All 96 affected individuals had a change of just one base in the junk. Instead of a C base, these patients had a G base. None of the relatives with the normal number of digits had a C in this position. Single base changes were also found in other families where some individuals had extra fingers. These changes were in the same general region of the genome as in the Dutch family but 200–300 base pairs away from that alteration.[2]

The junk region that carries these single base changes is an enhancer of the morphogen gene.* In order to create the correct body pattern, the spatial and temporal control of the morphogen is very tightly controlled by a whole slew of regulators. In the people with the mutation and the extra digit, the enhancer activity was slightly abnormal. The impact of the tiny change in this one regulator shows just how important and finely tuned this control is.

Here's some help with another quiz. What's the connection between Dutch people who have trouble buying gloves, and one of the great figures of 20th-century American literature? No? Give up? Well, in the 1930s Ernest Hemingway was given a cat by a ship's captain. Instead of having five toes on its front paws, this cat had six. There are now about 40 descendants of this cat at Hemingway's home, about half of whom have six toes on their front paws. It's easy to find pictures of these cats on the internet[3] and they are simultaneously cute and a little bit scary. The extra toe looks like a thumb, rendering the cats slightly too capable-looking for comfort.

The same group that identified the change in the enhancer region in humans with extra fingers showed that the same region was altered in Hemingway's cats. By inserting the enhancer

* The enhancer region is called ZRS and is found on the long arm of chromosome 7.

into another animal's genome they confirmed that the alteration changed the expression of the morphogen. The experimental animal over-expressed the morphogen and developed an extra digit on each front paw. Rather delightfully, this effect was demonstrated by inserting feline DNA into a murine embryo. A genuine cat-and-mouse game.[4]

Cats with extra front paw toes have also been found in other countries, including the UK. In the British cats there is also a change in the same enhancer, but it's not exactly the same change. It is two base pairs away from the Hemingway change, in a three-base-pair motif that is very highly conserved in evolution. The enhancer region that is involved in the extra digits on the forelimbs of humans and cats is about 800 base pairs in length and most of it is highly conserved from humans all the way down to fish. This suggests that the control of limb development is a very ancient system.

Morphogens and facial development

The morphogen that is responsible for finger formation is also critical for other developmental processes. One of these is the process whereby the structures of the front of the brain and the face are formed. If this process goes wrong, the effect can be very mild: simply a cleft lip. But at the other extreme, where the morphogen expression is more severely disrupted, the effects can be devastating. The brain and face may be completely abnormal, with no proper formation of brain structures. In the most severe cases the babies are born with just one malformed eye in the middle of the forehead and with severely impaired brain development. The babies never survive.

This spectrum of condition is known as holoprosencephaly.[5] A number of different protein-coding genes has been shown to be mutated in different families with this condition. Many of these genes are involved in the regulation of the same morphogen that is required for correct digit formation. In some cases, the gene for the morphogen protein itself is mutated. The developing embryo only produces half of the normal amount of the morphogen, because the functional protein is only produced from one chromosome, not two. The abnormalities in the affected individuals show that it is critical that the morphogen levels hit the right thresholds at key points in development.

Not all the mutations that cause holoprosencephaly have been identified. Researchers studied the DNA from nearly 500 individuals who were affected by the condition. They found an unexpected change in a junk DNA region of one severely affected infant. This was a single base change, from C to T, in a region over 450,000 base pairs away from the morphogen gene.[6]

The C to T change occurred in a block of ten base pairs that has been conserved since our ancestors diverged from the ancestors of frogs, over 350 million years ago. We can therefore surmise that this stretch of apparent junk has been maintained throughout evolution and has a function. In the case of this specific enhancer, the C binds a transcription factor protein.* Transcription factors are the proteins which are unusual because they recognise specific DNA sequences, usually in promoters, and bind to them. Binding of transcription factors to a promoter is essential for switching on a gene. The key transcription factor for this enhancer can bind to the ten-base-pair motif when the DNA contains a C in the appropriate position, but not when it contains a T.

This change from a C to a T in the enhancer wasn't present in 450 unrelated healthy control individuals. That might make it seem very likely that this change was the cause of the problems in the patient, but it's important to remember that it was also only seen once in about the same number of patients with the condition. The baby's mother was unaffected, and as expected she had

* This transcription factor is called Six3.

the normal C base on both her chromosomes. But unexpectedly, the baby's father had the same genetic sequence at the enhancer as his child. One chromosome had a C at the relevant position and the other had a T in the same place. But the father was completely unaffected by any symptoms of holoprosencephaly.

Although this might seem like strong evidence against a role for this C to T change, the situation isn't that straightforward. In holoprosencephaly it's quite common that there are lots of differences in a family, even where the mutation that causes the symptoms is in the morphogen gene itself. Up to 30 per cent of the family members with the mutation have no symptoms at all, and in others the symptoms may vary a lot from person to person. The first situation is known as variable penetrance, and the second is referred to as variable expressivity.

Unfortunately, these are classic cases where biologists identify a phenomenon, give it a fancy technical name and then stop thinking about it. These phrases are used to describe the phenomenon but we forget that we really don't understand why it is happening. It's a fascinating area that remains poorly understood. It's possible that there are other subtle sequence variations in the genome that compensate for the effects of the DNA change in some people. This could include other enhancers working more strongly, and boosting expression of the morphogen. There may also be epigenetic compensation in some people, which nudges the expression of key genes in a certain direction. It may be a combination of both these factors, plus others that we have not yet identified.

But where we have this uncertainty – parent and child with the same genetic change but different symptoms – it's vital to develop additional lines of evidence to support any hypothesis about the impact of the variant base. The researchers who identified the C to T change in the enhancer did exactly this, by testing the effect of this change in a mouse model. They showed that when the C was present, this stretch of junk DNA acted as an enhancer of morphogen expression. But when the C was replaced by a T the region no longer acted as an enhancer, and levels of the morphogen never reached the critical levels in the brain.

Morphogens and the pancreas

The morphogen that is implicated in the development of extra digits or in the various forms of holoprosencephaly isn't the only example of a human condition caused by a change in a regulatory region of DNA. There is a condition known as pancreatic agenesis, in which the pancreas fails to develop properly. Babies born with this condition often have severe diabetes.[7] This is because the pancreas is the organ that produces insulin, the hormone that allows us to regulate the levels of sugar in our blood.

The majority of families with pancreatic agenesis have a mutation in one particular transcription factor*,[8] but in a small number of affected families a different transcription factor is involved.**,[9] However, there are many cases where a child is born with unexplained pancreatic agenesis, when no one else in the family has ever been affected. Normally we might think that these cases have appeared randomly, perhaps as the consequence of something going wrong in development in response to an unidentified stimulus in the environment. But it became clear that most of these apparently sporadic cases occurred in families where the parents of the affected child were related to each other, typically cousins. This is known as consanguinity. When consanguinity is associated with higher rates of a disorder, we normally look for a genetic change. The change will be one where both copies of a chromosome carry the same variation. The reason why these conditions are more common in consanguineous couples is shown in Figure 15.1.

* This transcription factor is called GATA6.

** This transcription factor is called PTF1A.

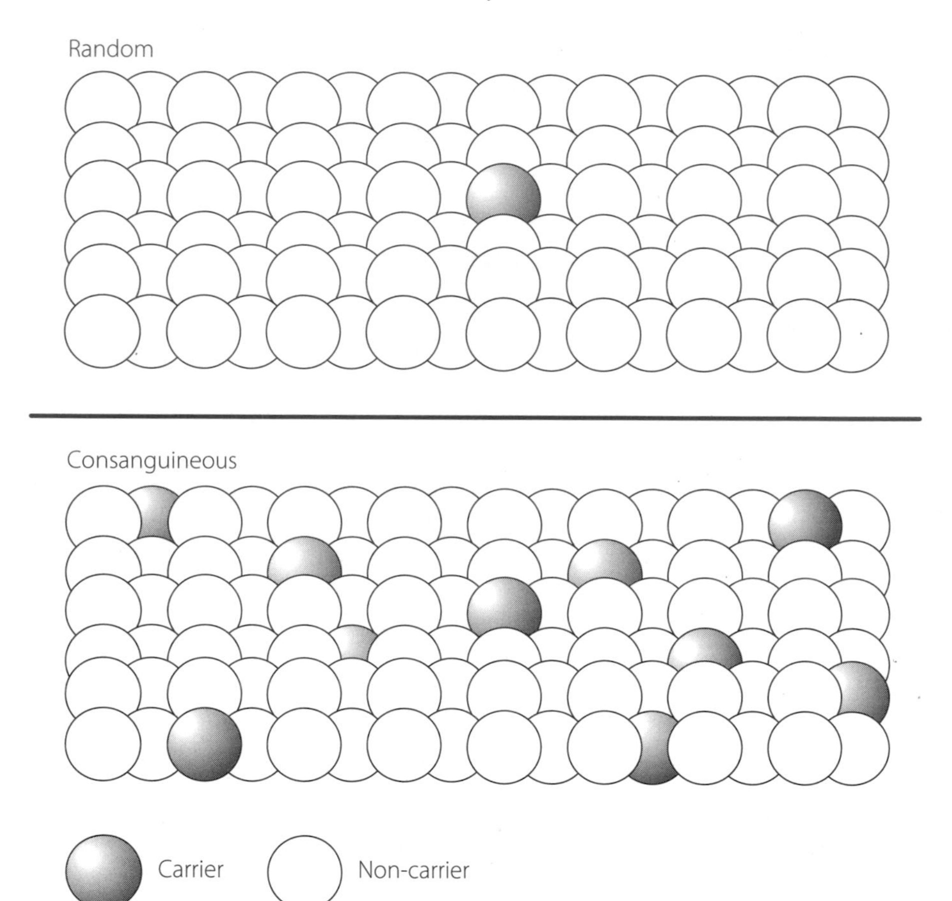

Figure 15.1 The upper part of the figure shows how a person who carries a rare genetic mutation is statistically relatively unlikely to meet another person with the same mutation in the general population. In their own family, however, it is much more likely that someone else will also have inherited the same mutation – a situation illustrated in the lower part of the figure. This is why rare recessive disorders (where the parents are asymptomatic carriers with one mutant gene each) present more commonly when the parents are related, for example first cousins.

Researchers isolated DNA from patients with the sporadic form of pancreatic agenesis and analysed all the protein-coding regions. They were unable to find any variations in sequence that could explain the disease. So they turned their attention to predicted regulatory regions. There are, as we have seen, an awful lot of predicted regulatory regions in the human genome. In order to narrow down their search, the investigators studied what happens when stem cells differentiate into pancreas cells in culture. They looked for regulatory regions which carried epigenetic modifications normally associated with enhancer function, and which bound transcription factor proteins known to be important in the development of pancreas cells.

This narrowed the list of candidate regions to just over 6,000, a much more manageable number to analyse in depth. Four patients each had a change from an A to a G in a putative enhancer region of about 400 base pairs on chromosome 10. This region lay 25,000 base pairs away from one of the transcription factors that is known to be mutated in a small number of families with pancreatic agenesis. Seven out of ten unrelated patients all had the same change: the enhancer on both copies of chromosome 10 had a G where there is normally an A base. Two patients had other nearby mutations, and the tenth patient had lost the enhancer altogether. Nearly 400 unaffected people were analysed. None of them carried this A to G change.

The researchers showed experimentally that the region they had identified acted as an enhancer in developing pancreatic cells, and also showed that the region loses its enhancer activity when the A is changed to a G. In further experiments, they explored how this enhancer regulates its target gene. This is shown in Figure 15.2. Briefly, the enhancer loops out so that it lies close to its target gene. The enhancer normally binds transcription factors which help to switch on the target gene. But transcription factors only bind to certain DNA sequences. When the A is changed to a G the transcription factors can't bind, and so they can't switch on their target gene.[10]

It's a bit like going fishing. Drop a hook baited with a juicy worm into the lake, and a carnivorous fish will bite. Drop in a

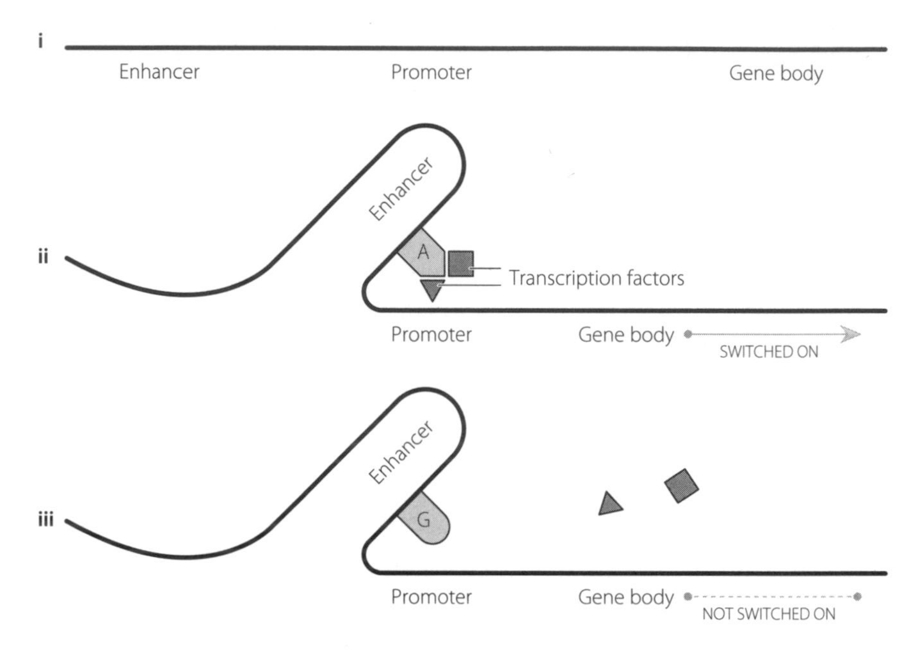

Figure 15.2 i shows the sequential positioning of an enhancer, promoter and gene body. In ii the DNA folds, bringing the enhancer close to the promoter. When the enhancer contains an A base in a specific position, the enhancer can bind specific proteins called transcription factors. These can activate the promoter and switch on the gene. In iii the A base in the enhancer is replaced by a G base, and the transcription factors can't bind. This in turn means that they can't activate the promoter, and the gene isn't switched on.

hook baited with carrot and there won't be so much as a nibble. Everything else is the same – the hook, the line, the sinker, the fish. But changing just one critical component (the bait) dramatically alters the chances of a successful catch.

Variations on a theme

It would be tempting to think that alterations in regions of junk DNA, which actually turn out to be regulatory regions, all have horrible consequences for the cell and the person. But that's because it's sometimes easier to look at abnormal situations than normal ones. This is especially the case if we are assessing the difference between having a disease and being healthy. In the cases described above, the single base changes in the regulatory regions have had dramatic effects. But these types of variations are also responsible for situations which are much less binary, and are just a normal part of human diversity.

Consider pigmentation. Pigmentation is a complex trait, by which we mean it is influenced by lots of genes acting together. The end results in this case are eye, hair and skin colour. We all know by experience that humans vary enormously with respect to these features of our appearance. In addition to several genes contributing to pigmentation levels, there are also different variants of those genes, creating additional potential for variation.[11]

One of the major variants is a single base difference, which occurs as either a C or a T. The T version is associated with higher levels of dark pigment, the C version with lower levels.* But this variation doesn't lie in a protein-coding gene. It has been shown that the reason it affects pigmentation is because it is in an enhancer region, 21,000 base pairs away from the target gene. This target gene codes for a protein that is important for pigment production. We know this because mutations in this gene lead to a form of albinism where the affected individual can't make pigment.[12,]**

It has been shown experimentally that the enhancer loops to the target. Transcription factors that control the target bind with greater or lesser efficiency depending on the C or T base.[13] This is very similar to the situation outlined above for pancreatic agenesis, and uses pretty much the same mechanism as shown in Figure 15.2.

It's quite likely that there are a lot of similar relationships between single base changes in junk DNA and the expression of protein-coding genes. This has implications for understanding

* This variant base pair has the catchy name of rs12913932.

** This gene is called OCA2.

human diversity and human health and disease. There are a large number of conditions where we know genetics plays a role in whether or not a person develops a disorder. In these conditions, a person's genetic background influences their likelihood of suffering from an illness, but doesn't explain it entirely. The environment also plays a role; as, sometimes, does plain bad luck.

We can identify disorders with a genetic contribution by looking at how often a disease occurs in a family. Twins are particularly useful in this analysis. Let's look at Huntington's disease, a devastating neurological disorder caused by a mutation in one gene. If one twin has the condition, their identical twin will also always have the disease (unless they die early from an unrelated cause, such as a traffic accident). Huntington's disease is 100 per cent due to genetics.

But if we look instead at schizophrenia, we find that if one twin suffers from this condition, there is only a 50 per cent chance that their identical twin is also affected. This has been calculated by analysing lots of twin pairs and working out the frequency with which both twins develop the condition. This tells us that genetics contributes about half the risk for developing schizophrenia and the other risk factors aren't due to the genome.

Researchers can extend these studies into other family members, because we know how much genetic information family members share. For example, non-identical siblings share 50 per cent of their genetic information, as do parents and children. First cousins share only 12.5 per cent of their genomes. It's possible to use this information to calculate the contribution of genetics to a large range of conditions from rheumatoid arthritis to diabetes, and from multiple sclerosis to Alzheimer's disease. In these conditions, and many more, genetics and environment act together.

If we can find enough families, we can analyse their genomes to identify regions that are associated with disease. But we have to remember that the data we generate will be very different from the simple situation we see with a purely genetic condition such as Huntington's disease. In Huntington's 100 per cent of the genetic contribution lies in one mutation in one protein-coding gene. But for a condition such as schizophrenia, the 50 per cent genetic contribution to the disease isn't due to just one gene, and the same is true of most other conditions where both genetics and the environment play a role. There could be five genes each contributing 10 per cent of the risk for schizophrenia, or twenty genes each contributing 2.5 per cent of the risk. Or any other combination of which you can think. This makes it harder to identify the relevant genetic factors, and to prove that sequence changes really do influence the condition being studied.

Notwithstanding these difficulties, more than 80 diseases and traits have been mapped using these methods, generating thousands of candidate regions and variations.[*,14] Remarkably, nearly 90 per cent of the regions identified across these studies are in junk DNA. About half are in the regions between genes, and the other half are in the junk regions within genes.[15]

Guilt by association

We have to be very careful about assuming that because we can detect a variation in DNA that is associated with disease, that this means the variation has a role in causing the disease. Sometimes we may just be looking at guilt by association. The genetic change that really contributes to the condition may be a different variation that is close by, and our candidate may just have been carried along for the ride.

An example of guilt by association would be cirrhosis of the liver. One way to assess exposure to cigarette smoke is to measure

* This approach to finding disease/trait-associated genes and variants is known as GWAS – genome-wide association studies.

the levels of carbon monoxide in a person's breath. Ten years ago, if we measured the levels of this gas in the breath of non-smokers with liver disease we might find that there were higher concentrations of this gas in the airways of people with the condition than without, on average. One interpretation (although not the only one) would be that passive smoking increases the risk of liver cirrhosis. But in reality, the carbon monoxide levels are a case of guilt by association. They probably just reflect that the patient may spend a lot of time in pubs and bars, because excessive alcohol consumption is a major risk factor for developing this illness. Until the introduction of smoking bans in many cities, pubs and bars were traditionally pretty smoke-filled environments.

Even if we exclude guilt by association when analysing the contribution of a genetic variation to a human disorder, we still need to be really careful to test hypotheses about the functional consequences of our findings. Otherwise, we can be badly misled.

The variation that contributes to human pigmentation that we met earlier in this chapter actually lies in the introns, the bits of junk DNA that lie in between the protein-coding parts of a gene, which we first met in Chapter 2. The gene is very big, and the variant base pair is in the 86th stretch of junk DNA between amino acid-coding regions. But this gene itself plays no role in control of pigment levels. So we have clear precedent for accepting that variations in the junk regions in one gene may be important for effects on other genes.

Obesity is one area in which there has been a great deal of interest in identifying genetic variants linked to physical variation. Nearly 80 different regions in the human genome have been associated with obesity or with other relevant parameters such as body mass index.[16]

In multiple studies, the variation showing the greatest association with obesity was a single base pair change in a candidate protein-coding gene on chromosome 16.*[,17,18] Individuals who inherited an A on both copies of this gene tended to be about 3kg (6.6lb) heavier than individuals who inherited a T on both copies. This change was in the junk region between the first two amino acid-coding stretches of the candidate gene. The fact that this association was detected in more than one study was important as this increases our confidence that we are looking at a meaningful event.

It seemed that a consistent story was developing, because experiments in mice seemed to confirm a role for this gene in control of body weight. Mice that were genetically manipulated so that they over-expressed this gene were overweight, and developed type 2 diabetes symptoms when they ate a high-fat diet.[19] When this gene was knocked out in mice, the animals had less fat tissue and a leaner body type than control mice. Even when the knockout mice ate a lot, they burnt loads of calories, even though they weren't particularly active.[20]

This created a lot of excitement. It implied that if scientists could find a way to inhibit the activity of this gene in humans, they might be able to develop an anti-obesity drug. There was still a problem because we aren't altogether sure what the candidate gene does in cells, and that makes it difficult to create good drugs. But at least we had a starting point. Both the human and mouse data implied that the gene coded for an important protein in obesity and metabolism. This was coupled with the reasonable assumption that the variant base pair associated with obesity affected the expression of the gene itself.

But in the immortal words of Mitch Henessey, the character played by Samuel L. Jackson in *The Long Kiss Goodnight*, 'Assumption makes an ass out of "u" and … "umption".' Of course, hindsight is always 20/20 and there's no reason to feel

* This gene is called FTO or 'fat mass and obesity associated'.

condescending to the scientists who were exploring the role of the protein. It's just that nature seems to have a way of tripping us up.

Here's the real reason why that single base pair variation makes a difference to human physiology. There's another protein-coding gene half a million base pairs away from the key single base pair change described above.* The junk region of the original gene interacts with the promoter of the second gene, altering its expression patterns. Essentially, the junk region acts as an enhancer. The effect is seen in humans, mice and fish, suggesting it is an ancient and important interaction.

The investigators looked at the expression levels of this second gene in over 150 human brain samples. There was a clear correlation between the base pair variant in the junk/enhancer region and the expression levels of the second gene. But there was no correlation between the base pair variant and the expression levels of the original candidate, the gene that actually contains the variation.

When the researchers knocked out expression of the second gene in mice they found that, compared with control animals, the mice were lean, had low adipose tissues and increased baseline metabolic rate. This was in fact the first time anyone realised that the second gene was involved in metabolism at all.[21]

What we have here is a model very similar to the one we have already encountered for human pigmentation and for pancreatic agenesis. There are in fact a number of different variant base pairs in the junk region of the original obesity-associated gene. Many of these have been associated with obesity. This suggests that all of these variants probably have the same effect, i.e. they change the activity of the enhancer, and thereby alter the expression levels of the target gene, half a million base pairs away.

Of course, the mouse data suggest that the original gene, the one that contains the variations in its junk DNA, may also play

* This gene is called IRX3, or Iroquois homeobox protein 3.

a role in obesity and metabolism. So we could ask if, in practical terms, it really matters how the single base pair changes bring about their effects. But there is a way in which this matters a lot, and that's in the field of drug discovery.

One of the many problems in developing new drugs is that frequently some patients will respond to a drug and others won't. This adds a lot of additional expense. It means that pharmaceutical companies have to run very large clinical trials to see if their drug works, because they have to test it in all-comers. It also means that it's expensive to use the drug in clinical practice, because the doctor will give it to all patients with the relevant condition, but it will only work in some of them.

These days, pharmaceutical companies are all trying to create something called 'personalised medicine'. This means that they try to develop drugs for situations where they know very early on which patients they want to treat, usually based on their genetic background. This can be very effective. It means drugs cost less money to develop, are usually licensed faster, and are only given to patients who are likely to benefit. This is an advantage for the health care providers because they aren't wasting money treating people who won't respond. It's also potentially better for the patients, as all drugs have possible side effects, and there's no point having the risk of side effects if there is very little likelihood of receiving benefit.[22] There have been real successes in this approach, most notably in drugs for breast cancer,[23] a blood cancer[24] and most recently lung cancer.[25]

The critical step in developing personalised medicines is to identify a reliable biomarker. The biomarker tells you which of your potential patients should respond to your drug. Ideally, you want a situation where 100 per cent of people with the relevant biomarker will respond to the drug. The problems start if you have the right biomarker for the disease, but you link it with the wrong target. You will create a drug and then be stuck wondering why patients who 'should' respond, don't. It will be because there is a break in the circle of relationships, as shown in Figure 15.3.

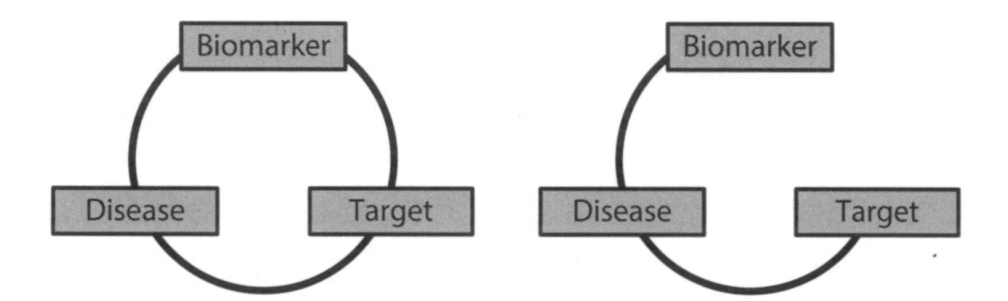

Figure 15.3 On the left-hand side there is a perfect relationship between the biomarker, target and disease. On the right-hand side there is no relationship between the target and the presence or absence of the particular biomarker. Under these conditions the biomarker is useless for predicting which patients with the disease will respond to a drug developed against the target.

The potential market for obesity drugs is, with no pun intended, huge. It's likely some companies had already started drug discovery programmes against the original target, which they will now be either terminating, or trying to find a way to salvage. In the meantime, portion control and a bit of exercise remain our best bet.

16. Lost in Untranslation

There are few crimes lower than deliberately hurting a child. In many countries, staff in emergency departments are trained to look for patterns of unexplained injuries including fractures in babies and toddlers. Often such a medical history will result in children being taken into care, little or no parental access, and ultimately prosecution and possibly imprisonment of one or both parents.

Protection of a child is of course paramount. But imagine the nightmare for parents if this happens to them and they are entirely innocent, because the fractures are due to an undetected medical condition.[1] Although the number of such miscarriages of justice is small compared with genuine cases of child abuse, the effects for the family are devastating. Loss of liberty, marital breakdown, social exclusion and, most heartbreakingly, the loss of parent–child contact.

A genetic condition can and has led to this misdiagnosis of child battery on more than one occasion. The disorder is called osteogenesis imperfecta, but it's more commonly known as brittle bone disease.[2] Patients with brittle bone disease suffer fractures very easily, sometimes from mild traumas that might not even cause much of a bruise in a healthy child. The same bones may break repeatedly, and they may heal imperfectly, so that the affected person becomes increasingly disabled over time.

We might think that this condition should be very recognisable, making it rather strange that parents are sometimes wrongly accused of hurting their children. But there are a number of factors that complicate the picture. The first is that brittle bone disease affects about six or seven children in 100,000. A doctor may simply have never encountered the condition, especially if they are relatively new to emergency medicine. But sadly, they probably will encounter child battery and so are more likely to have this as a default diagnosis.

The diagnosis is also complicated because there are at least eight different types of brittle bone disease, varying in their severity and the fine details of the presentation. At the most extreme, babies may suffer fractures even before they are born. The different forms of brittle bone disease are caused by mutations in different genes. The most common ones are defects in collagens, proteins that are important for making sure bones are flexible. Although we often think of bones as very rigid, it's important that they have some flexibility, so that they bend rather than break in response to movement. It's the same principle behind why we teach children not to climb on dead trees, because the inflexible, dried-out branches are more likely to break than the green, bendy limbs of living trees.

In most cases of brittle bone disease, only one copy of a gene is mutated. The other copy (because we inherit a copy from each parent) is fine. But having one normal copy isn't enough to compensate for the effects of the 'bad' gene. Usually when this happens we expect to see a disorder in not just the child, but also in one parent. This is the parent who passes on the condition to their baby. But if the mutation is a new one, created during the production of eggs or sperm, a child can be affected without their parent having any symptoms. This tends to be particularly the case in the very severe forms of brittle bone disease. This makes it harder for doctors in an emergency room to recognise that they are looking at a condition caused by a mutation.

But if doctors do suspect that a baby may be suffering from brittle bone disease, they can order genetic tests to try to confirm their diagnosis. The genetic diagnosis will involve analysing the sequences of the genes that are known to be mutated in brittle bone disease. Scientists will prioritise the order in which they sequence the genes by looking at the details of the patient's symptoms, and deciding which form of brittle bone disease they think they have. Then they'll sequence the most likely genes first, looking for mutations that alter the proteins required for strong healthy bones.

This usually works well. But inevitably we find that there are some patients with all the symptoms of brittle bone disease but who don't have any mutations that alter the amino acid sequence in the proteins known to be involved in this condition. This is exactly the situation that faced scientists trying to understand the cause of a specific class of brittle bone disease* in a small number of Korean families. In this class of cases, there are characteristic patterns of fractures, but also a very strange after-effect. When the bones are damaged, either by a fracture itself or by medical intervention to repair a break, the patient's body responds in an unusual way. It lays down too much calcium around the injury site, creating an obvious cloudy effect visible on an X-ray.

At the same time, other researchers were analysing a child from a German family, who had the same highly unusual type of brittle bone disease. Remarkably, the cases in both Korea and Germany were caused by exactly the same mutation. Just one base pair among the 3 billion the affected children inherited from each parent was altered. And the alteration that caused this disease was not in the amino acid-causing region of a gene. It was in junk DNA.

The beginning and the end

The mutation lay in a region of junk we have already encountered. In Chapter 2, we saw how protein-coding genes are composed

* Osteogenesis imperfecta type 5.

of modules. The modules are initially all copied into messenger RNA and various modules are joined together. Regions that don't code for protein are removed during this 'splicing' process (*see page 16*).

But two regions of junk DNA always remain in the mature messenger RNA. These were shown in Figure 2.5 and Figure 16.1 depicts them again. Because these regions at the beginning and end of the messenger RNA are retained but never translated into protein, they are known as untranslated regions.* Although they don't contribute to the amino acid sequence of the normal protein, researchers are identifying new ways in which these untranslated regions contribute to protein expression and to human health and disease.

The researchers in Korea analysed the DNA sequences of nineteen patients. Thirteen of these came from three affected families,

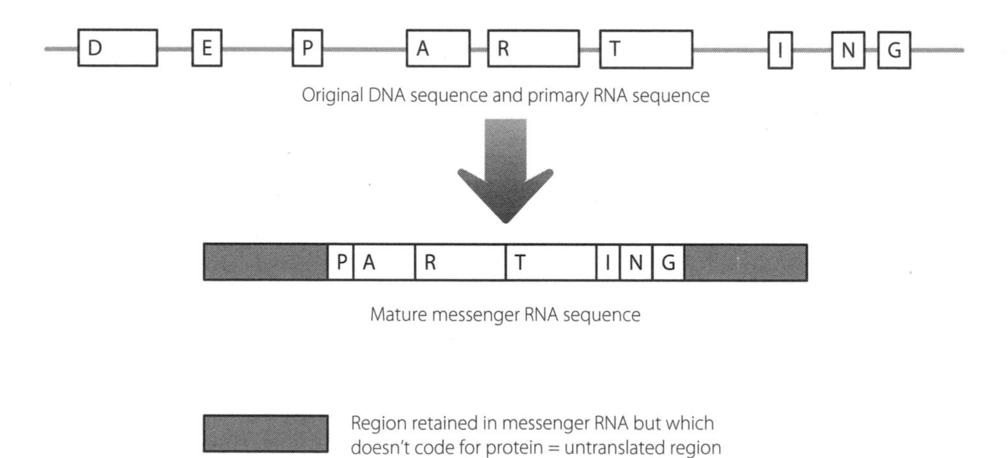

Figure 16.1 Even after the amino acid-coding regions of a messenger RNA have been spliced together, there is still some junk RNA which is retained in the molecule, at the beginning and end.

* These are often referred to in the literature as UTRs, for untranslated regions. The one at the beginning of the messenger RNA is called the 5′UTR and the one at the end of the messenger RNA is called the 3′UTR.

and the other six were single cases. Each of the nineteen patients had a change from a C base to a T base in the untranslated region at the start of the protein-coding region of a specific gene.* This change was just fourteen bases away from the start of the protein-coding region of the messenger RNA. They didn't detect this C to T change in any of the unaffected family members or in 200 unrelated people from the same ethnic background.[3]

At about the same time, the researchers 5,000 miles away in Germany found exactly the same mutation in a young girl with the same type of brittle bone disease, and in another unrelated patient. In both cases, it was a fresh mutation. It wasn't present in the parents and must have arisen during the production of eggs or sperm.[4] The scientists analysed the same region of the genome from over 5,000 unaffected people and found no one with this change.

There is a bit of a puzzle when we look at our image of messenger RNA in Figure 16.1. In the diagram the protein-coding regions and the untranslated regions have been drawn so that they look different from each other. But this isn't what they are like in the cell. In reality, they look the same at the sequence level, because they are just formed from RNA bases.

For anyone fluent in written English, the following is pretty easy to decipher:

Iwanderedlonelyasacloud

Even though all the letters have been run together, we can recognise where individual words start and stop. The same is true for the cell, which is able to tell the difference between the sequences in the untranslated regions and in the amino acid-coding regions of a messenger RNA.

Translation of messenger RNA to create protein is carried out

* The gene is called IFITM5.

at the ribosomes, in a process that we met in Chapter 11. The messenger RNA is fed through the ribosome, starting at the beginning of the messenger RNA molecule. Nothing much happens until the ribosome reads a particular three-base sequence, AUG (as mentioned in Chapter 2, the T base in DNA is always replaced by a slightly different base called U in RNA). This signals to the ribosome that it's time to start joining up amino acids to create a protein.

Using our example from above, it would be as if we looked at a piece of text that read as:

dbfuwjrueahuwstqhwIwanderedlonelyasacloud

The capital I acts as the signal to us to start reading proper words, fulfilling a similar purpose to the AUG that signals the start of translation.

In the genes of the Korean and German patients with brittle bone disease, there is a point at which the normal DNA sequence in the untranslated region changes from ACG to ATG (which will be AUG in RNA). The consequence is that the ribosomes start the protein chain too early. This is shown in Figure 16.2.

This results in a strange phenomenon where junk RNA is changed to protein-coding RNA. This adds an extra five amino acids to the start of the normal protein, as shown in Figure 16.3. The protein involved in this type of brittle bone disease is one that has parts inside and outside the cell. The alteration in the junk DNA adds an extra five amino acids to a part of the protein that is outside the cell.

It's not quite clear why these five amino acids cause the symptoms of the disease. Previous experiments in rodents had shown that too much or too little of this protein leads to defects in the skeleton, so it's clear that having exactly the right amount of the protein is important.[5] The extra five amino acids are on a part of

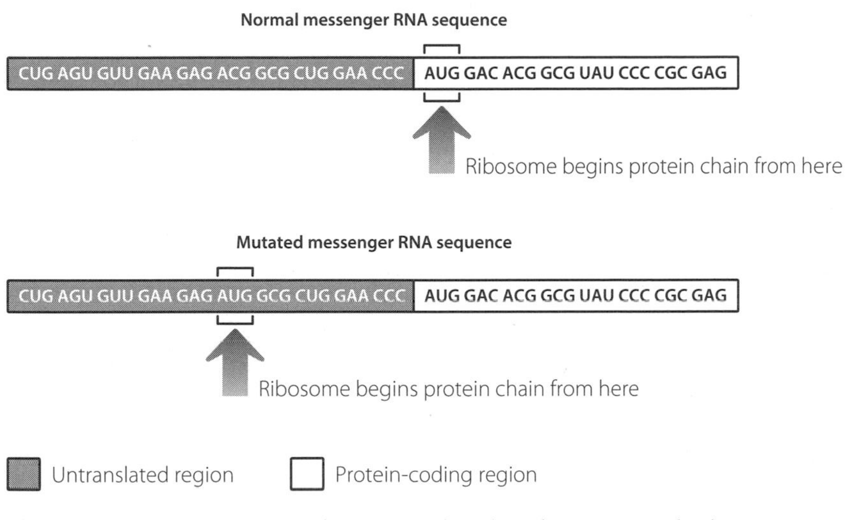

Figure 16.2 A mutation in the untranslated junk region at the beginning of the messenger RNA mis-directs the ribosome. The ribosome begins sticking amino acids together too early, creating a protein with an extraneous sequence at the beginning.

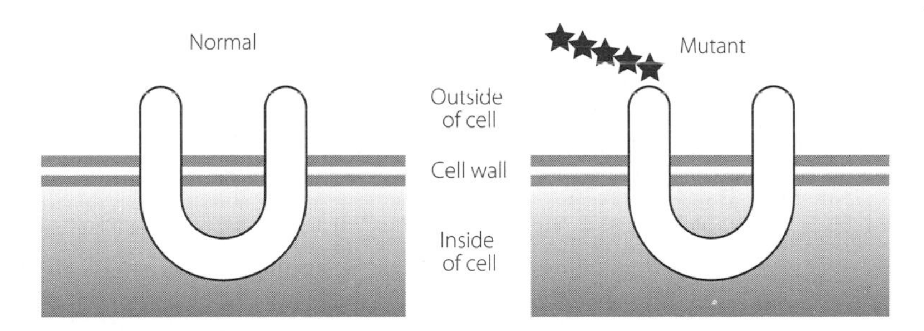

Figure 16.3 The U-shaped protein on the right has an extra five amino acids at the beginning, represented by stars. These extra amino acids probably influence which other molecules can interact with this protein.

the protein that we would expect might bind to other proteins or molecules that signal to the bone cells. It may be that having these extra five amino acids stops the mutant protein from responding properly, like putting chewing gum on the sensor of a smoke detector.

Brittle bone disease isn't the only human disorder caused by mutations in the untranslated regions at the start of a gene. There is a strong genetic component in about 10 per cent of cases of melanoma, the aggressive skin cancer. A mutation has been identified in some of these genetically driven cases that works in a very similar way to the problem in brittle bone disease. Essentially, a single base change in the untranslated region at the start of a gene creates an abnormal AUG signal in the messenger RNA. This again results in the ribosome starting the amino acid chain too early in the gene sequence. This creates a protein with extra amino acids at the start, which behaves in an abnormal way, increasing the chances of cancer.[6]

As always, we need to beware of seeing patterns from too little data. Not all mutations in the untranslated region at the start of a gene create new amino acid sequences. There is another type of skin cancer which is usually much less aggressive than melanoma. This is called basal cell carcinoma, and it too has a strong genetic component. A rare mutation was found in a father and his daughter, both of whom developed this kind of tumour.

The untranslated region at the start of a particular gene usually contains the sequence CGG, repeated seven times, one after the other. The affected father and child had an extra copy of the CGG. Having eight repeats rather than seven predisposed them to basal cell carcinomas. This mutation didn't change the amino acid sequence of the protein encoded by the gene. Instead, the extra three bases seemed to change the way the messenger RNA was handled by the ribosome, in ways that aren't very clear. The end result was that the cells of the patients expressed much less of the specific protein than normal.[7]

Cancer is a multi-step disease, and although these mutations in the untranslated region at the start of certain genes predisposed the patients to tumours, other events probably also took place in the cells before full-blown cancer developed.

In the beginning was the mutation

But we have already encountered a disorder where an inherited mutation in the untranslated region at the start of a gene leads directly to pathology. This is the Fragile X syndrome of mental retardation (*see page 19*). As a reminder, the mutation is an unusual one. A three-base-pair sequence of CCG is repeated far more times than it should be. Anything up to 50 copies of this repeat is considered to be in the normal range. Fifty to 200 copies is not normally associated with disease, but once the number of repeats gets into this range it becomes very unstable. The machinery that copies DNA for cell division seems to have trouble keeping count of the number of repeats, and even more repeats get added. If this happens in the gametes, the resulting child may have many hundreds or even thousands of the repeats in their gene, and they present with the Fragile X syndrome.[8]

The longer the repeat, the lower the expression of the Fragile X gene. As we saw in an earlier chapter, this is because of cross-talk with the epigenetic system (*see page 123*). Where C is followed by G in our genome, the C can have a small modification added to it. This is most likely to happen in regions where this CG motif is present at high concentrations. The large number of CCG repeats in the Fragile X expansion provide exactly this environment. The untranslated region in front of the Fragile X region becomes very highly modified in the patients, and this switches the gene off. Fragile X patients don't produce any messenger RNA from this gene, and consequently don't produce any protein from it either.

The effects on the patient of this lack of protein are dramatic. Patients are intellectually disabled but also have symptoms reminiscent of some aspects of autism, including problems with social interactions. Some patients are hyperactive, and some suffer from seizures.

This of course makes us wonder what the protein normally does. The clinical presentation is quite complex, which suggests that the protein is probably involved in complicated pathways, and this indeed seems to be the case.

As we saw in Chapter 2, the Fragile X protein is usually complexed with RNA molecules in the brain. The protein targets about 4 per cent of the messenger RNA molecules expressed by the neurons.[9] When it binds these messenger RNA molecules, the Fragile X protein acts as a brake on their translation into proteins. It prevents the ribosomes from producing too many protein molecules from the messenger RNA information.[10]

This extra level of control on gene expression seems to be particularly important in the brain. The brain is an extraordinarily complex organ, and the cell type that is of most interest to us is the neuron. This is what people usually mean when they talk about brain cells. There are an awful lot of neurons in the human brain, the most recent estimate being just over 85 billion.[11] Each brain contains twelve times as many neurons as there are people on earth. And in the same way that people have complex networks of friends, acquaintances, lovers, families and enemies, neurons are also linked in. What's startling is the degree of connection between the billions of neurons. Neurons send out projections that connect with other neurons in vast networks, constantly influencing each other's responses and activities. The precise number of connections is really difficult to estimate, but each cell probably makes at least 1,000 connections with other neurons, meaning our brains contain at least 85 trillion different contact points.[12] It makes Facebook look positively parochial.

Establishing these contacts appropriately is a huge task in the brain. Think of it as arranging to see good friends frequently while trying to avoid the weird guy you met in your first week at college. Contacts are set up and then either strengthened or pruned back, in complex responses to environment and to activities of other neurons in the network. Many of the target messenger RNAs that bind to the Fragile X protein under normal conditions are involved in maintaining the plasticity of the neurons, allowing them to strengthen and prune connections as appropriate.[13] If the Fragile X protein isn't expressed, the target messenger RNAs are translated into protein too efficiently. This messes up the normal plasticity of the neurons, leading to the neurological problems seen in the patients.

Researchers have recently shown that they can use this information to treat Fragile X syndrome, at least in genetically engineered animals. Mice which lack the Fragile X protein have problems with their spatial memory, and with their social interactions. A mouse that can't find its way around and doesn't know how to react to its fellow mice is a rodent that won't last long. Researchers used these mice and applied genetic techniques to dial down the expression of one of the key messenger RNAs that would normally be controlled by the Fragile X protein. When they did this, the scientists detected marked improvements in the animals. Spatial memory was better and the mice behaved appropriately around other mice. They were also less susceptible to seizures than the standard Fragile X mouse models.

These symptomatic improvements were consistent with underlying changes that the scientists detected in the brains of the animals.[14] Neurons in normal brains have little mushroom-shaped spines that are characteristic of strong, mature connections. The neurons of humans and mice with Fragile X syndrome have fewer of these, and a larger number of long, spindly, immature connections. After the genetic treatment, there were more mushrooms and fewer noodles.

The most exciting aspect of this was that it suggested it could be possible to improve neuronal function even after symptoms had developed. We can't use the genetic approach in humans but these data imply that it is worth trying to find drugs that will have a similar effect, as a potential means of treating Fragile X patients. This syndrome is the commonest inherited form of mental retardation so the benefits of developing a treatment could be dramatic both for individuals and for society.

Now for the other end

As we saw at the start of this book, expansions in a three-base sequence at the other end of a gene can also cause a human genetic disease. The best-known example is myotonic dystrophy, which is caused by expansion of a CTG repeat in the untranslated region at the end of a gene. Repeats of 35 units or above are associated with disease, and the larger the repeat, the more severe the symptoms.[15]

Myotonic dystrophy is an example of a gain-of-function mutation. The main effect of the expansion in the Fragile X gene is to stop production of its messenger RNA. But this isn't the case in myotonic dystrophy. The mutant version of the myotonic dystrophy gene is switched on, resulting in messenger RNA molecules with large expansions at the end of the molecule. It's these multiple copies of CUG in the messenger RNA (remember that T is replaced by U in RNA) that cause the symptoms. If we turn back to Figure 2.6 (*see page 23*), we can see in outline how this happens. The expanded repeats act like a molecular sponge, soaking up particular proteins that are able to bind to them.

Junk DNA plays a remarkable role in myotonic dystrophy, as shown in Figure 16.4. The CTG expansion in the junk untranslated region binds abnormally large quantities of a key protein.* This protein is normally involved in removing the junk DNA that is found between amino acid-coding regions when DNA is first copied into RNA. Because so much of the protein is sequestered onto the expanded myotonic dystrophy untranslated repeat, it can't carry out its normal function very well. Consequently, lots of RNA molecules from different genes aren't properly regulated.

---

* This protein is called Muscleblind-like protein 1, or MBNL1.

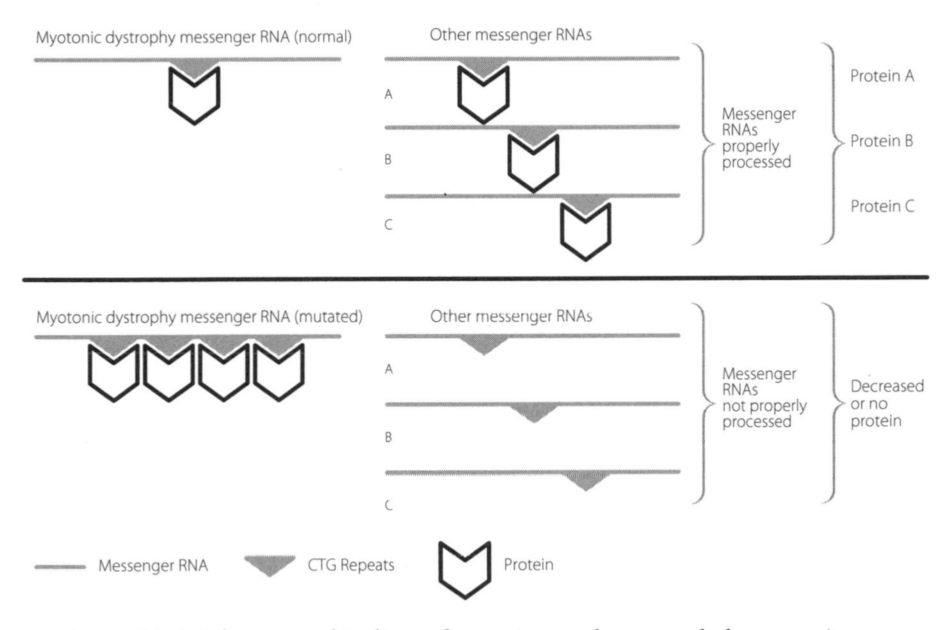

Figure 16.4 The excess binding of proteins to the expanded myotonic dystrophy repeat in the messenger RNA sequesters the proteins away from other RNA molecules that they should also be controlling. The other messenger RNAs are no longer properly processed, and this disrupts production of the proteins that they should be used to produce.

This titration of the binding protein, which occurs in any tissues where both it and the myotonic dystrophy gene are expressed, plays a large role in explaining why the disease can present so differently in different patients. Instead of being all-or-nothing, varying proportions of the binding protein may be 'left over' to regulate its target genes. The proportion will depend on the size of the expansion and the relative amounts of myotonic dystrophy messenger RNA and binding protein in a cell.[16]

It is worth looking in a bit more detail at the proteins that are ultimately affected by these deficits (proteins A, B and C in Figure 16.4). The best-validated ones are the insulin receptor,[17] a heart protein[18] and a protein in skeletal muscle that transports chloride ions across membranes.[19] Insulin is required to maintain muscle mass. If the muscle cells don't express enough of the receptor that binds insulin, they will start to waste away. The heart protein is one that we know is important for the correct electrical properties of the heart.[20] Transport of chloride ions across skeletal muscle membranes is an important stage in the cycles of muscle contraction and relaxation. So, the defects in the processing of the messenger RNAs coding for these proteins are consistent with some of the major symptoms in myotonic dystrophy, i.e. muscle wasting, sudden cardiac death because of fatal abnormalities in heart rhythm, and the difficulty in relaxing a muscle after it has contracted.

Myotonic dystrophy is a great example of the importance of junk DNA in human health and disease. Although the mutation lies in the messenger RNA produced from a protein-coding gene, the mutation has little if any effect on the protein itself. Instead, the mutated RNA region is itself the pathological agent, and it causes disease by altering how the junk regions of other messenger RNAs are processed.

Say 'AAAAAAAAA'

The untranslated regions at the end of protein-coding messenger RNAs have a number of functions in normal circumstances. One of the most important involves a process that affects all messenger RNA molecules. 'Naked' messenger RNA molecules can be broken down in a cell very quickly, via a process that probably evolved to help us get rid of certain types of viruses rapidly. In order to stop this happening, and to make sure the messenger RNA molecules linger long enough to be translated into protein, the messenger molecules are modified very soon after production. Essentially, lots of A bases are added to the end of the messenger RNA, by a process that is outlined in Figure 16.5. There are usually about 250 A bases on the end of a mammalian messenger RNA. They are important for stability and also for making sure that the messenger

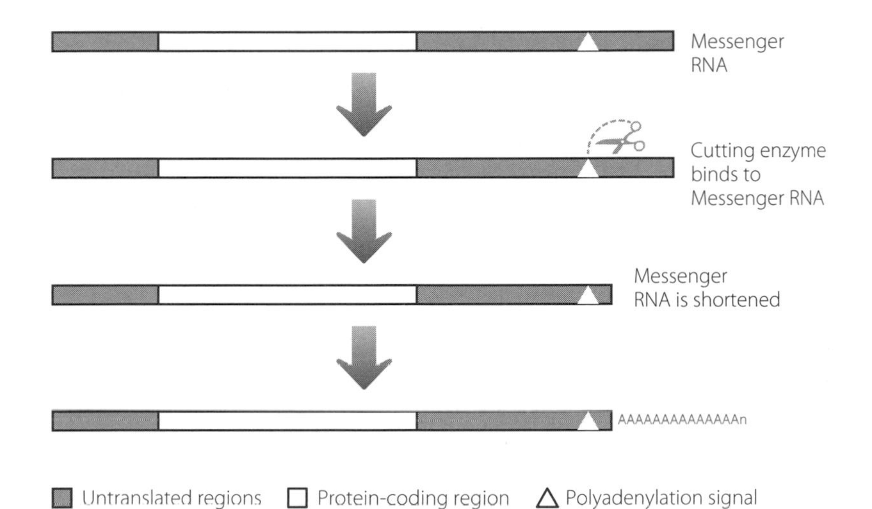

Figure 16.5 A sequence in the untranslated region at the end of a messenger RNA attracts an enzyme (shown by the scissors) that binds at a specific site and then cuts the molecule a little further along. Lots of A bases are added to the cut end of the messenger RNA molecule, even though these were not coded for in the original DNA sequence.

RNA is exported out of the nucleus where it is made and into the ribosomes where it is translated into protein.

There is a critical motif in the untranslated region at the end of the messenger RNA. This is shown by the triangle in Figure 16.5 and is called the polyadenylation signal (the A base is adenosine, so adding lots of A bases is called polyadenylation). This is a sequence of six bases (AAUAAA) within the junk of the untranslated region. It acts as a signal for a messenger RNA-processing enzyme. The enzyme recognises the six-base motif, and cuts the messenger RNA a little distance away, usually ten to 30 bases further downstream. Once the messenger RNA has been cut in this way, another enzyme can add the multiple A bases.*

* This is known as a non-templated change because there is no underlying DNA template for these A bases in the genome.

This six-base motif often occurs many times in the same untranslated region. It's not particularly clear how a cell 'chooses' which motif to use at any one time. It is probably influenced by other factors in the cell. But because there are multiple motifs that can be used, there may be multiple messenger RNAs that code for exactly the same protein, but which contain different lengths of the untranslated region before the multiple As. These different-length messenger RNAs will have different stabilities and so produce different amounts of protein from each other. This creates additional opportunity for fine-tuning the amount of protein that is produced.[21]

There's a very unusual genetic condition in humans called IPEX syndrome.* It's a fatal autoimmune disease in which the body attacks and destroys its own tissues. Cells lining the intestine are attacked, resulting in severe diarrhoea in young infants and a failure to thrive. The glands that produce hormones can also be attacked, leading to conditions that include type 1 diabetes, where patients can't produce insulin. The thyroid gland may also be targeted, resulting in underactivity.[22]

Rare cases of IPEX syndrome are caused by a mutation in the polyadenylation signal. Instead of the normal AAUAAA sequence, there is a single base change. As a consequence, the six-base sequence becomes AAUGAA and no longer acts as a target for the cutting enzyme.[23]

The gene where this change occurs codes for a protein that switches on other genes.** This protein is required to control a particular type of immune cell.*** In some genes the change in a single six-base motif might not be that serious a problem, because the cell would use other, nearby, normal six-base sequences in the same

* IPEX stands for Immunodysregulation, Polyendocrinopathy, Enteropathy, X-linked.

** FOXP3, a transcription factor.

*** Regulatory T cells.

untranslated region. This might disrupt fine-tuning a little, but we wouldn't expect to see anything as severe as IPEX syndrome. The problem arises in IPEX because the untranslated region of this gene contains hardly any other suitable six-base motifs to act as signals for polyadenylation. The mutation in the untranslated region means that the messenger RNA isn't cut properly, A bases aren't added and the messenger RNA is very unstable. Because of this, the cells produce hardly any of this protein. Essentially, the effects of the mutation in this junk motif are as bad as if the protein-coding region itself had been disrupted.

It's only fairly recently, as sequencing technologies have become cheaper, that researchers have really started analysing the untranslated regions of messenger RNA molecules to identify mutations that cause rare instances of serious diseases. We can be pretty confident that over the next few years we will see many more examples of this. One of the reasons we can be bullish about this prediction is that researchers may have already identified another such example.

Amyotrophic lateral sclerosis (ALS), also known as motor neuron disease or Lou Gehrig's disease, is a devastating disorder. Neurons in the brain and spinal cord which control muscle movement die off progressively. Sufferers become increasingly wasted and paralysed, unable to talk, swallow or breathe properly.[24] The cosmologist Stephen Hawking suffers from ALS, although his case is rather atypical. He was first diagnosed at the age of 21, whereas most people with ALS develop their first symptoms in middle age. Professor Hawking has survived for over 50 years with the condition, but sadly most patients die within five years of diagnosis, although this period may be increasing with better medical intervention.

There is much that we still don't understand about ALS. Less than 10 per cent of cases run in families. In the other 90 per cent there may be variations in DNA that predispose someone to the condition if they encounter environmental triggers (which we can't yet identify). Some patients may also have a mutation that is sufficient on its own to cause the condition, even without a family history of the disorder. This mutation may have arisen in the eggs or sperm of their parents, for example.[25]

One of the genes involved in ALS is believed to be responsible for 4 per cent of cases that run in families, and 1 per cent of cases that occur without a family history.[*,26,27,28] In all the original cases involving this gene, the mutations were in the protein-coding regions. Researchers have now identified four different variants in the untranslated region at the end of this gene. These were found in patients with ALS who didn't have any other known mutations. Although these could just be harmless variations, the distribution of the protein and its expression levels were abnormal in the cells from these patients. These findings are at least suggestive that the changes in the untranslated region led to abnormalities in the processing and translation of the protein itself, leading to disease.[29]

* The gene is called FUS – Fused in Sarcoma.

17. Why LEGO is Better Than Airfix

Most children, and quite a few adults, enjoy making models. There are various ways of doing this but let's just look at the extremes. One of the most popular formats in the UK for over 30 years was the Airfix kit. Small plastic parts specific to an aircraft, ship, tank or just about anything else you can think of (Bengal Lancer, anyone?) were supplied with detailed instructions. The user glued the parts together, painted them, applied transfers and admired the finished article for years after.

At the other extreme is that universal Danish toy of which I am so fond, LEGO. Although there are lots of specialist LEGO kits now, the concept remains the same as ever. A relatively limited number of components that can be joined together in any combination the user wants. And the model can always be split back down into its original bricks and reused to create something else.

Simple organisms like bacteria tend more to the Airfix way of life. Their genes are fairly set, coding for just one protein. The more complex an organism becomes, the more the genome begins to resemble LEGO, with a much greater degree of flexibility in how the components are used. And when we think how extraordinary we humans are, it seems reasonable to say, in a nod to a certain movie, that at the genomic level, 'everything is awesome'.*

An extreme version of this phenomenon is the splicing that our cells can use to create multiple related proteins from one gene, as

* If you haven't already, go and see *The Lego Movie*, it's excellent.

in Figure 2.5 (*page 18*). This ability to use the components of a gene in multiple different ways creates enormous flexibility and added opportunities for an organism. We can get some idea of the amount of variability that is possible by looking at some of the numbers involved. Human genes contain an average of eight amino acid-coding regions, each separated by an intervening stretch of junk DNA.* At least 70 per cent of human genes have been shown to create at least two proteins.[1] This is achieved by joining up different amino acid-coding stretches. Using our DEPARTING example (as in Figure 2.5), this allows us to produce the protein DART but also the protein TIN. The ability to create different proteins this way is known as alternative splicing.

The regions that code for amino acids are short compared with the intervening junk regions. The stretches that code for amino acids have an average length of about 140 base pairs, but they can be surrounded by junk regions that are several thousand base pairs in length.[2] About 90 per cent of the base pairs in a gene are from the intervening sequences, not from the amino acid-coding ones. If we think of this in terms of the English language, this immediately shows us some of the problems the cell faces.

Imagine you meet someone and are extraordinarily smitten by them. You have heard that they love poetry, and you want to sweep them off their feet, but you always skipped literature class in school. A friend gives you a piece of paper with a killer first line from a poem on it. But for some reason your vaguely sociopathic friend has split up the words of the first line among a load of gibberish, and you only have a couple of seconds to find the poetry, say it out loud and win someone's heart (or at least their attention). Can you do it? Take a very quick look at Figure 17.1 and find out.

This is what our cells do all the time, every moment of every

* As a reminder, the junk regions between amino acid-coding regions are known as introns. The amino acid-coding parts themselves are known as exons.

lqrrtliruienvjbhghadbwnfqwrhvierhbtuehufjebjxmbmvnkbnvmnnlehaboiwhebrijjjoovburunvrmwwmwuhtyghdlsqppjfn
bjcbbvfxkmxmsfdhdhjfkmjmljllgnhjwekvfdhbutfjvnytuututriobvbvmcncnmzxmciiwerbfnjcxegnxwcbeihfcnzihxbhnzxmx
kmjvbecgfvbchvgcbfdncmxkmazkjcfhcbnxzkxcfbvworldfbcdnxszmxcjhgbvfcnhadxxncvfcxszxchcahfgevbgbuhruhtieiyuo
yttirqrutiopqwieueoiwpvbkvbncmzxmxcbnvskdkjfhgfdgueriwruytreiwohfghjxncbnvnxcmzncbvjfhgfjdskafgeriowuryteri
owiurghjfkdnbvncmxncbvnxmcznbcnmxcnbfghjerguitaroeiwuytirohgfkdlsxmcdkemcknjbhbhuvdmkmxwokszlpazqaqlxp
dceofvingnkmokokokkokkonbvcxxcfvcxzcrxcyfvmgbmvncbxvbdcnmvbhmoibnuvevxbencmorvbmbnvcxbnmcvbnvucxnj
bvnjcdiwbcndiwbhnjfnbhvnjnnfdbhubhcudebhvbhncjnbnhjitokmkyojnbgovfnjchduxsvgtfcrfwvgdbuehrnbtkmbkvfmndi
uhvswfdvhugnhkhongefhdvydefghtnjhjkmkimjoenoughkhtgnjfdewbrkjum,imojhgijrfbdwsfraxeswwzexrdxsessxdxdxdrc
xdrcdcfcfcftgvbyhnmkmplkmjhyugthkyhljukhgfrdefrngmbhnmhbvdxbdntocmvgbngvfdxsbnmfvgbhomgvfdbwnxfjghun
gvfijunjcefhubhnrgijthniewdhubhnfjrijbnjiehrbhntjigvfnjdewhfbnjfrunijbdehfurbgbugjnfeidjwncdkmwokxnicdefjgrubh
ubfrhdwsbhuxsncidfergijhgbufhewdydrinkvsgdfbibhnbjvifdcbhndijfandvnjokcdsnqjuhdvfgyhudbcijwnmokmcdokfvmob
ghmnokjmknhkbgmrfdjwinshuwbgvtfcdxftcdbuvfjmkfmnvjdbcdfbgkfdnhcdvtimefghrufncdsoibcvhufbdjnvjgbijvfbdchh
bchvjncdoxnoksmocnivifcndnicdnicdnvfnjvfncmxxmxmxnuyuyfjdnmoqwhufhyrgyehduhequmjpufruifrubdjbuhcnuher

A glorious line of poetry is in here somewhere……

Figure 17.1 Have a quick look, can you find the line that will win someone's heart?

day of our lives. Machinery in the cell analyses a long stretch of apparent gibberish and almost instantaneously finds the hidden words and joins them together. You can take a look at Figure 17.2 to see if you managed to compete with the non-sentient proteins that keep you alive.

In any long stretch of random letters there will also be combinations that spell words just by chance. Use these words by mistake when wooing (does anyone still woo?) the object of your desires and you may ruin your one chance for happiness. Figure 17.3 will show you how.

lqrrtliruienvjbhg**had**bwnfqwrhvierhbtuehufjebjxmbmvnkbnvmnnlehaboiwhebrijjjoovburunvrmwwmwuhtyghdlsqppjfn
bjcbbvfxkmxmsfdhdhjfkmjmljllgnhj**we**kvfdh**but**fjvnytuututriobvbvmcncnmzxmciiwerbfnjcxegnxwcbeihfcnzihxbhnzxmx
kmjvbecgfvbchvgcbfdncmxkmazkjcfhcbnxzkxcfbv**world**fbcdnxszmxcjhgbvfcnhadxxncvfcxszxchcahfgevbgbuhruhtieiyuo
yttirqrutiopqwieueoiwpvbkvbncmzxmxcbnvskdkjfhgfdgueriwruytreiwohfghjxncbnvnxcmzncbvjfhgfjdskafgeriowuryteri
owiurghjfkdnbvncmxncbvnxmcznbcnmxcnbfghjerguitaroeiwuytirohgfkdlsxmcdkemcknjbhbhuvdmkmxwokszlpazqaqlxp
dceofvingnkmokokokkokkonbvcxxcfvcxzcrxcyfvmgbmvncbxvbdcnmvbhmoibnuvevxbencmorvbmbnvcxbnmcvbnvucxnj
bvnjcdiwbcndiwbhnjfnbhvnjnnfdbhubhcudebhvbhncjnbnhjitokmkyojnbgovfnjchduxsvgtfcrfwvgdbuehrnbtkmbkvfmndi
uhvswfdvhugnhkhongefhdvydefghtnjhjkmkimjo**enough**khtgnjfdewbrkjum,imojhgijrfbdwsfraxeswwzexrdxsessxdxdxdrc
xdrcdcfcfcftgvbyhnmkmplkmjhyugthkyhljukhgfrdefrngmbhnmhbvdxbdntocmvgbngvfdxsbnmfvgbhomgvfdbwnxfjghun
gvfijunjcefhubhnrgijthniewdhubhnfjrijbnjiehrbhntjigvfnjdewhfbnjfrunijbdehfurbgbugjnfeidjwncdkmwokxnicdefjgrubh
ubfrhdwsbhuxsncidfergijhgbufhewdydrinkvsgdfbibhnbjvifdcbhndijf**and**vnjokcdsnqjuhdvfgyhudbcijwnmokmcdokfvmob
ghmnokjmknhkbgmrfdjwinshuwbgvtfcdxftcdbuvfjmkfmnvjdbcdfbgkfdnhcdv**time**fghrufncdsoibcvhufbdjnvjgbijvfbdchh
bchvjncdoxnoksmocnivifcndnicdnicdnvfnjvfncmxxmxmxnuyuyfjdnmoqwhufhyrgyehduhequmjpufruifrubdjbuhcnuher

One of the most romantic and seductive first lines of poetry in the English language.
"Had we but world enough and time" from Andrew Marvell's ***To His Coy Mistress***

Figure 17.2 The words shown in **bold and underlined** should do the trick for you.

lqrrtliruienvjbhg**had**bwnfqwrhvierhbtuehufjebjxmbmvnkbnvmnnlehaboiwhebrijjjoovburunvrmwwmwuhtyghdlsqppjfn
bjcbbvfxkmxmsfdhdhjfkmjmljllgnhj**we**kvfdh**but**fjvnytuututriobvbvmcncnmzxmciiwerbfnjcxegnxwcbeihfcnzihxbhnzxmx
kmjvbecgfvbchvgcbfdncmxkmazkjcfhcbnxzkxcfbv**world**fbcdnxszmxcjhgbvfcn*had*xxncvfcxszxchcahfgevbgbuhruhtieiyuo
yttirqrutiopqwieueoiwpvbkvbncmzxmxcbnvskdkjfhgfdgueriwruytreiwohfghjxncbnvnxcmzncbvjfhgfjdskafgeriowuryteri
owiurghjfkdnbvncmxncbvnxmcznbcnmxcnbfghjerguitaroeiwuytirohgfkdlsxmcdkemcknjbhbhuvdmkmxwokszlpazqaqlxp
dceofvingnkmokokokkokkonbvcxxcfvcxzcrxcyfvmgbmvncbxvbdcnmvbhmoibnuvevxbencmorvbmbnvcxbnmcvbnvucxnj
bvnjcdiwbcndiwbhnjfnbhvnjnnfdbhubhcudebhvbhncjnbnhjitokmkyojnbgovfnjchduxsvgtfcrfwvgdbuehrnbtkmbkvfmndi
uhvswfdvhugnhkhongefhdvydefghtnjhjkmkimjo**enough**khtgnjfdewbrkjum,imojhgijrfbdwsfraxeswwzexrdxsessxdxdxdrc
xdrcdcfcfcftgvbyhnmkmplkmjhyugthkyhljukhgfrdefrngmbhnmhbvdxbdn*to*cmvgbngvfdxsbnmfvgbhomgvfdbwnxfjghun
gvfijunjcefhubhnrgijthniewdhubhnfjrijbnjiehrbhntjigvfnjdewhfbnjfrunijbdehfurbgbugjnfeidjwncdkmwokxnicdefjgrubh
ubfrhdwsbhuxsncidfergijhgbufhewdy*drink*vsgdfbibhnbjvifdcbhndijf**and**vnjokcdsnqjuhdvfgyhudbcijwnmokmcdokfvmob
ghmnokjmknhkbgmrfdjwinshuwbgvtfcdxftcdbuvfjmkfmnvjdbcdfbgkfdnhcdv**time**fghrufncdsoibcvhufbdjnvjgbijvfbdchh
bchvjncdoxnoksmocnivifcndnicdnicdnvfnjvfncmxxmxmxnuyuyfjdnmoqwhufhyrgyehduhequmjpufruifrubdjbuhcnuher

If a combination of the **right** and *wrong* words is selected,
the sentiment may be very different e.g. "Had we but had enough to drink".

Figure 17.3 No! Bad combination!

By using this slightly bizarre example, we can understand some of the mechanistic challenges that our cells face when splicing RNA molecules properly. If we were designing this as a process, it would have the components shown in Figure 17.4.[3] In addition to the components described in this diagram, it's important to realise that different cells will handle the same gene differently, depending on the cell type and what is happening to it at any given moment. Consequently, all the stages have to be appropriately regulated and integrated so that the correct protein variants are made to meet the needs of the situation.

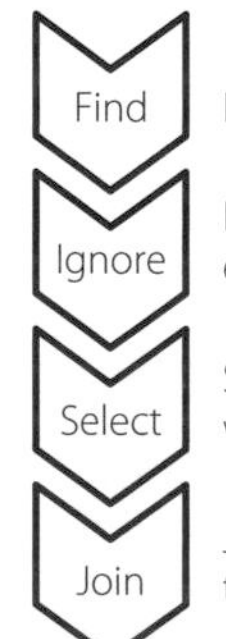

Identify regions that code for amino acids

Ignore regions that don't code for amino acids, even if they look like they might (pseudo-regions)

Select the amino acid-coding regions that will be joined together

Join the right regions together to create messenger RNA

Figure 17.4 The sequence, reading from the top, lays out the steps that the splicing machinery has to be able to carry out to join up the appropriate amino acid-coding regions to create the correct mature messenger RNA.

The splice of life

This splicing of long RNAs to create smaller messenger RNAs that carry the information for specific proteins is a really complex process. It's a very ancient system, and the components and steps have been maintained from yeast throughout the entire animal kingdom. It is carried out by a huge conglomeration of molecules called the spliceosome, which forms the splicing machinery. The spliceosome is composed of hundreds of proteins and also some junk RNAs, a little like the ribosomes that act as the factories to produce proteins.[4]

One of the critical stages is that the spliceosome wraps around the intervening sequences that need to be removed from an RNA molecule. It snips them out and then joins up the amino acid-coding regions. It's an enormously complicated multi-stage process but we know that one of the first key steps is that the spliceosome needs to recognise the intervening regions, so that it can bind to them and remove them.

The beginnings and ends of these intervening sequences are always indicated by particular two-base sequences. Junk RNA molecules in the spliceosome can bind to these two-base sequences in much the same way as the two strands of DNA can pair up in our genes.

But there are only four bases in RNA, which means there are only sixteen two-base sequences (AC and CA are considered as different pairs, as are all the others). We would expect that the two-base sequences that mark the beginnings and ends of the intervening sequences would also be found elsewhere in these sequences, and also in the amino acid-coding regions. This is indeed the case. So although these two-base sequences are necessary for splicing, they aren't sufficient on their own to direct the process properly. Other sequences are also required, as indicated in Figure 17.5.

The other sequences involved in selecting how splicing will take place are found in both the junk intervening regions and the

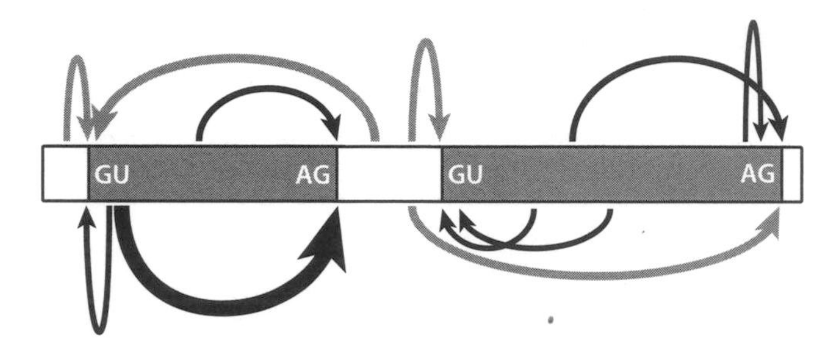

Figure 17.5 Multiple sequences within an RNA molecule interact to drive splicing. The two-base motifs shown are necessary but not in themselves sufficient to regulate all the fine-tuning of this process. Other sites are involved, of varying strengths, as indicated by the different sizes of arrows.

amino acid-coding regions. Some of them influence splicing very strongly, others are more subtle. Some increase the chances of a splice event, others decrease them. They work in complex partnerships and the impact that they have on the final splicing pattern is affected by other things happening in the cell, such as the precise complement of proteins in the spliceosome. The descriptions that are used for these modifying sequences usually include such words as 'dizzying' or 'bewildering'. These are geek speak for 'unbelievably complicated, way beyond anything we can get our heads around or even design predictive computer algorithms for at the moment.'

Splicing and disease

We can get clues to the degree of sophistication by looking at a group of genetic diseases. These include a form of blindness called retinitis pigmentosa, which affects about one in 4,000 people. The blindness is progressive, often starting in the teenage years with a decline in night vision, and then becoming steadily worse and more disabling with age. The loss of vision occurs because the cells in the eye that detect light gradually die off.[5] About one in twenty cases is caused by a mutation in one of five proteins involved in a specific step in splicing.[6,7,8,9] The mutation only causes a deficit in the cells of the retina, and not in all the other cells in the body which also rely on splicing. This shows us that splicing is under complex cell- and gene-specific control, in ways that we haven't yet been able to understand.

By contrast, there is a very severe form of dwarfism with other unusual features such as dry skin, sparse hair, seizures and learning disabilities. Affected children almost always die before they are four years old.[10] It's very rare except in the Ohio Amish community, where 8 per cent of the people are carriers. That's because the mutation that causes this condition was present in the small number of families that founded this community. It isn't found in other Amish groups such as those in Pennsylvania, which were founded by other families. When the mutation that causes this condition was identified, the researchers first thought that it was changing the amino acid sequence of a gene that codes for a splicing protein. But we now know that the change actually disrupts the three-dimensional structure of a junk RNA that forms part of the spliceosome.[11] Unlike the retinitis pigmentosa situation, this defect in the action of the spliceosome causes a very wide-ranging set of symptoms, possibly by causing mis-splicing of lots of different genes.

Human disorders don't just occur because of defects in the splicing machinery. They can also arise because protein-coding genes themselves have mutations in sites that are important for the control of splicing of the RNA from that single gene. Some authors have claimed that up to 10 per cent of human inherited disorders may be caused by mutations at the splice sites, those two-base sequences shown in Figure 17.5.[12]

One example of this mechanism was a family in which two young siblings developed intractable diarrhoea within a few days of birth. Medical staff managed to stabilise the children, but the diarrhoea persisted for many months and one of the two affected children died at seventeen months of age. When the genomes of the children were sequenced, the researchers found a mutation in a splice site in a gene, changing one of the GU sequences shown in Figure 17.5. This resulted in the splicing machinery skipping over an amino acid-coding region inappropriately. Essentially, an amino acid-coding region was left out of the protein, and as a consequence the protein could no longer do its job.[13]

Kaposi's sarcoma is a cancer that first came to public attention when it was found at high levels in people with AIDS. AIDS is caused by the human immunodeficiency virus (HIV) and the effect of the HIV infection is to suppress the immune system. Kaposi's sarcoma is caused by a different virus called HHV-8. Normally our immune systems control this virus but if the immune system is seriously below par, HHV-8 can become established and trigger Kaposi's sarcoma.

HHV-8 is present in a high percentage of people in the Mediterranean basin, but Kaposi's sarcoma is rare in this population, and almost never found in small children. So medics were very surprised when a Turkish family brought in their two-year-old daughter who had a classic lesion characteristic of this cancer on her lip. The cancer spread rapidly and aggressively and the little girl died just four months after she was first diagnosed.

The child was negative in all tests for HIV. Her parents were related to each other, a first-cousin marriage. Researchers looked for genetic reasons why the daughter might have an impaired immune response to HHV-8.

By sequencing DNA obtained from samples that had been taken from the deceased girl, scientists identified a mutation in a splice site of a specific gene. The mutation changed an AG to an AA, which meant the spliceosome could no longer recognise where it was meant to cut the RNA molecule. The result was that a junk region that should have been removed was retained in the messenger RNA molecule. This messed up the sequence, creating a stop signal much too early in the messenger RNA. This prevented the ribosome from making the full-length protein. Because the protein is one that is required for mounting a good immune response to viruses such as HHV-8, the child with the mutation was very susceptible to Kaposi's sarcoma.[14]

Although splice site mutations are relatively common, genetic diseases are more often caused by mutations in the amino acid-coding regions of genes. Some of these cause problems because they introduce stop signals that prevent the ribosomes from making full-length proteins from messenger RNA templates. Other mutations may change the code from one amino acid to another. For example, CAC codes for the amino acid histidine whereas CAG codes for glutamine, a different amino acid. But researchers have speculated that up to 25 per cent of the mutations that change the amino acid in this way also influence the splicing of nearby regions in the messenger RNA. In some cases the disease may be due not to the single amino acid alteration per se, but to the variation that the nucleotide change creates in the way a messenger RNA is spliced.

The problem is that it is very difficult to demonstrate that this is the case in most situations. Even if we can show that the change in the RNA leads to both an altered splicing pattern and an amino acid change, how can we tell which effect causes the disease symptoms? Are these due to protein with one altered amino acid, or because the protein has also been spliced in an unusual pattern?

Nature has actually provided us with proof that sometimes a mutation in a coding region can cause a disease by influencing splicing, rather than by changing an amino acid. There is an extraordinary disorder called Hutchinson-Gilford Progeria, named after the two scientists who first identified it. *Progeria* means early ageing and this particular form is incredibly dramatic. It is also extremely rare, affecting about one in 4 million children.[15]

Affected babies seem perfectly healthy at first but within a year their growth rate slows dramatically, and they remain underweight and short for the rest of their lives. The children begin developing many symptoms of old age, including thinning hair, stiffness and baldness. Although there are some ageing conditions that they don't develop, such as Alzheimer's disease (and the children also don't have learning disabilities), the affected individuals do develop severe cardiovascular disease. This is usually what causes death by the early teens, as a consequence of heart attacks or major strokes.

In 2003 researchers identified the gene mutation that causes Hutchinson-Gilford Progeria. Every patient they tested had a de novo mutation, meaning one that developed spontaneously in the parents' egg or sperm. Incredibly, in eighteen unrelated patients (out of twenty who were assessed) the mutation was exactly the same.[16]

A sequence that should read GGC in a particular gene had mutated and now read as GGT. This mutation was in the amino acid-coding part of the gene. This might seem like a straightforward case of a mutation changing an amino acid in a protein, so of course the first thing to do is to look at the genetic code and see what these two sequences code for. GGC, the normal sequence, codes for a simple amino acid called glycine. But the mutated sequence, GGT, codes for – wait for it – glycine. Yep, same amino acid.

This is because our genetic code has a level of redundancy. Our genome is composed of four letters – A, C, G and T (or U in RNA). Blocks of three letters are used to code for an amino acid. There are 64 possible combinations of three from four letters. Three of these combinations are stop signals, telling the ribosomes not to add any more amino acids to a protein chain. This leaves 61 combinations to code for amino acids. But our proteins only contain 20 different amino acids. So some amino acids can be coded for by different three-letter combinations. At one extreme, glycine is coded for by GGA, GGC, GGG and GGT(U). At the other, the amino acid methionine is only coded for by the combination AT(U)G.

But if the amino acid sequence encoded by the mutated gene doesn't change in Hutchinson-Gilford Progeria, what causes the dramatic phenotype in this condition? Look again at Figure 17.5. The two-base sequence at the beginning of each intervening junk region within a gene is GT. In the patients where the normal GGC changes to GGT, the amino acid region gains an inappropriate extra splice signal. In the context of all the other splicing signals in that genomic region, this inappropriately positioned GT acts strongly. The spliceosome cuts the messenger RNA in the amino acid-coding region rather than in the junk region. The amino acid-coding regions join up badly and the end result is a loss of about 50 amino acids from the end of the protein. This in turn means that the protein itself isn't processed properly, and it begins to wreak havoc in the cells. We still don't know exactly how this leads to the extraordinary ageing we see in these children, but our best guess at the moment is that the cell nucleus isn't maintained properly. This may lead to changes in gene expression and nuclear breakdown. Some genes and some cell types may be more sensitive to this than others.

There is another condition that affects young children called Spinal Muscular Atrophy. In this condition, the nerve cells supplying the muscles gradually die off, leading to muscle wasting and loss of mobility. There are a number of different forms, and in the most severe version the life expectancy for affected babies is very low, less than eighteen months.[17] It is relatively common for a genetic disease: in the UK about one in 40 people is a carrier, meaning about 1.5 million of us have one defective copy of the gene. Luckily, both copies of the gene have to be mutated for symptoms to develop.[18]

Spinal Muscular Atrophy is caused by the deletion or loss of function of a gene called SMN1. If we look at the human genome we might be surprised that this has such a major effect, because there is another gene that codes for exactly the same protein. This gene is called SMN2. This raises a really obvious question: since they code for the same protein, why can't the SMN2 gene compensate for a damaged/deleted SMN1 gene?

In a rather similar way to Hutchinson-Gilford Progeria, the SMN2 has one subtle variation from SMN1. This is a change in DNA sequence in an amino acid-coding region. It doesn't change the amino acid sequence, because it is in one of the three-base sequences with redundancy in terms of amino acid coding. Instead it changes one of the sites that help ribosomes to work out where to splice the messenger RNA molecule.[19] It doesn't change a splice site, it changes one of the sites that influences where splicing occurs. This results in skipping of an amino acid-coding region, and the production of a protein that isn't functional. Because of this, the SMN2 gene can't compensate for the malfunctioning SMN1 gene. The normal SMN1 protein is required for spliceosome activity. So essentially, a mutation in one gene leads to problems with overall splicing of messenger RNAs, and this could be overcome except for an independent splicing issue with a potentially compensatory gene.

Manipulating splicing for therapeutic gain

As we saw in Chapter 7, in Duchenne muscular dystrophy, the severe muscle wasting disease carried on the X chromosome, the dystrophin gene is mutated (*see page 92*). This gene is exceptionally large, stretching for almost 2.5 million base pairs. It contains nearly 80 amino acid-coding regions, which need to be spliced and processed properly. This is particularly important because the dystrophin protein is long-lasting. This means that any change that increases the chance of mis-splicing will affect the cell for a long time. But the presence of 78 introns in this massive gene means there is a high risk of spontaneous and inherited mutations which could affect splicing, just because there are so many opportunities for this to happen. As one review rather pithily puts it: 'The massive (2.4Mb) dystrophin gene, most of which is in its 78 introns, is a splicing accident waiting to happen, and it does so with an incidence of 1 in 3,000 live births.'[20]

So, some cases of Duchenne muscular dystrophy are caused by splicing defects. However, in a large number of cases the disease is caused because critical regions of the gene, and hence the protein, are missing. But in recent years there has been a glimmer of hope for developing treatments for this invariably fatal disease. Perhaps counterintuitively, this has been based on developing drugs which *encourage* abnormal splicing in the dystrophin gene in affected boys.

The dystrophin protein acts like a kind of shock absorber in muscle cells. We can think of the dystrophin molecules as being like the springs in a mattress. In order for the mattress to remain supportive, the springs need to attach to the top and bottom of the mattress. If there has been a manufacturing fault, and the springs have been produced without the last ten centimetres, they won't be able to attach to the top of the mattress. The more often you use the mattress, the less supportive and more distorted it will become.

It is quite common that Duchenne muscular dystrophy is caused by loss of internal regions of the dystrophin gene. When the gene is copied into RNA, the remaining regions are spliced together. Compared with the normal dystrophin gene, the mutant gene is now lacking some amino acids in the interior of the protein. But that's not what really causes the biggest problem, as shown in Figure 17.6.

The amino acid code is read in blocks of three bases, as we have already seen. When the correct amino acid-coding regions (known as exons) join together, as in the normal gene, they produce a long messenger RNA molecule that codes for lots of amino acids. But if wrong exons join together, they may be out of sync with each other, so that the blocks of three don't run properly. A really simple example would be the following:

YOU MAY NOT SEE THE END BUT TRY

If we lose one letter, we rapidly start to lose the sense:

YOU MAY OTS EET HEE NDB UTT RY

This is called a frame shift. In messenger RNA the first impact is that the wrong amino acids are inserted into the growing protein chain. But quite soon, something even more dramatic happens. There will be a combination of three letters that act as a stop signal. At that point the ribosomes will cease adding amino acids, and the mutated protein is truncated.

This is what happens in the patients with deletions of certain regions of the dystrophin gene. In Figure 17.6, the frame of reading the three-base combination is indicated by the numbers under the boxes. As long as the number at the end of one box and the beginning of the next add up to three, the ribosome can keep reading the messenger RNA. But where the most common deletion occurs, this introduces a frame shift, which rapidly leads to a stop signal and a severely shortened protein chain.

One way around this would be to encourage the cell to skip one of the amino acid-coding regions after the deletion, because this would restore everything to the right reading frame. The end result would be a protein that has a bit missing internally, but can still function reasonably well. This might slow down progression of symptoms. This is shown using our bed spring analogy, in Figure 17.7. The dystrophin molecule will still be able to connect to the necessary proteins at either end. It won't be quite as good

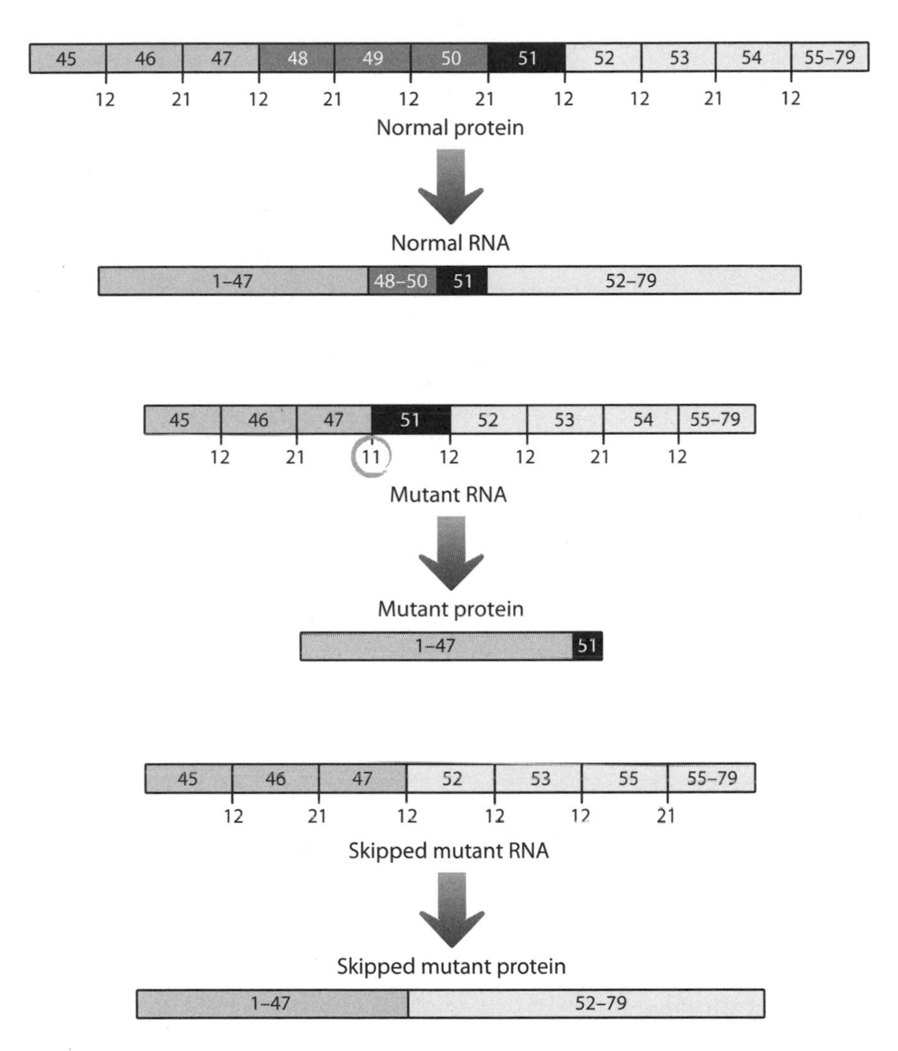

Figure 17.6 Representation of a key region where a mutation in the dystrophin gene can lead to a severely shortened protein molecule because of a shift in amino acid reading patterns when amino acid-coding regions 48 to 50 are lost from the DNA. In order to maintain the reading pattern, the numbers under each boundary must add up to three. If region 51 could be skipped in the mutant gene, the reading sequence would be restored. For simplicity, all the amino acid-coding regions have been drawn as the same size, although in reality they vary from one another.

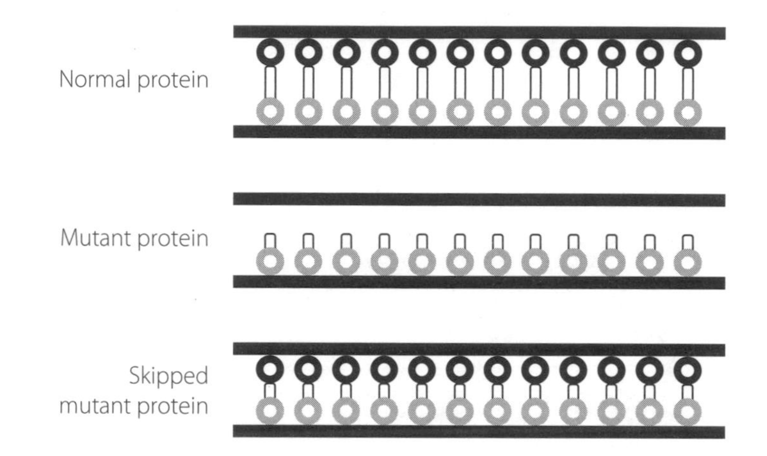

Figure 17.7 Schematic showing how a mutant dystrophin protein is unable to attach to the two sides of the cell membrane. The skipped version of the mutant protein, which is missing some internal sequences, can attach to the two sides of the membrane. Because it is shorter it isn't as good a shock absorber as the normal protein, but it is much better than the original mutant.

at absorbing shocks as the full-length protein. But it will be a lot better than one that can't tether to the necessary cellular structures.

The supporting evidence for this hypothesis looked good, and biotech companies began programmes to try to find a way of exploiting this knowledge. A company called Prosensa developed a drug that helped muscle cells to skip over amino acid-coding region 51 and eventually licensed this experimental drug to the pharmaceutical giant GlaxoSmithKline. In April 2013, GlaxoSmithKline published the results of a small trial in boys with the relevant form of Duchenne muscular dystrophy. Fifty-three boys were randomly assigned to two groups. One group received the drug, the other went through exactly the same procedures, but without actually receiving any of the experimental drug. This is known as a placebo procedure and is an important way of controlling for effects in clinical trials that are due not to the drug but to other effects such as increased optimism or patients getting better independently of the drug. The boys were tested 24 and 48 weeks later. The test measured how far they could walk in six minutes.

After 24 weeks, the boys who received placebo had got worse, as we would expect for this disease. They couldn't walk as far as when they entered the trial. But the boys who received the drug could walk more than 30 metres further than when the trial started. The boys were tested again after 48 weeks. The placebo group had deteriorated even more. In the six-minute walking test, their performance was almost 25 metres less than when the trial started. The boys who had been treated could walk over eleven metres further than when the programme began.[21]

These data showed that, over time, even the boys who received the drug began to decline (look at the difference between 24 and 48 weeks), but this decline was dramatically slower than when the condition was running its normal course.

The results from this trial caused enormous excitement. Finally, it looked like there might be hope developing in the treatment of a previously intractable disorder. Even if the treatment didn't cure the patients, it might significantly slow down the development of the irreversible symptoms. This was what everyone researching in the field, and the families of affected boys, had been working towards for decades. True, it wouldn't work for all Duchenne sufferers, but between 10 and 15 per cent of patients were expected to be eligible for this approach, based on the kind of mutation in their dystrophin gene.

Just six months later, those hopes were in tatters. GlaxoSmithKline ran a larger trial and this time couldn't find any significant difference between the treated and untreated groups.[22] The results from larger trials are more reliable than ones from smaller studies because they are less likely to be affected by odd patterns that look like a response but aren't. GlaxoSmithKline had no doubts about its large trial, convinced that if there had been a genuine effect of the drug, it would have been detected. They handed the drug back to Prosensa and walked away. Prosensa is continuing with clinical studies, although its share price tanked after GlaxoSmithKline departed, reflecting concerns by analysts that this programme may be doomed.

There is another company that is also trying to exploit splicing patterns to leap over the troublesome region in the dystrophin gene in the same patient groups. This company is called Sarepta, and it is using a similar approach to treating the affected boys. Although the company remains very upbeat about its programme, the Food and Drug Administration has questioned whether its trials are large enough to give genuinely conclusive results. For example, one of the studies in which a dramatic difference between the untreated and treated groups was seen only contained twelve patients.

Investors in the companies are no doubt feeling a chill breeze, but it can't begin to compare with what the families of affected boys must have gone through and be going through every day.

It would be tempting to look at the science in this chapter and decide that splicing is more trouble than it's worth. It certainly seems to be an example of Sod's law – if something can go wrong, it will. But the reality is that the same is true of almost every biological process. Billions of bases, thousands of genes, trillions of cells, billions of people. It's a numbers game; nothing goes right every time. But the fact that this process of joining together split genes has been maintained through hundreds of millions of years of evolutionary history, using a highly conserved system, makes it pretty clear that the advantages of the sophistication, additional information content and sheer flexibility more than compensate for the off days.

18. Mini Can Be Mighty

Perhaps because we are quite large animals, we tend to be most impressed by other large animals. And that's OK. After all, a big cat such as a jaguar is an impressive creature. We also tend to be impressed because the jaguar is a hunter, a top carnivore. An ant, by comparison, looks rather puny, even if it's one of the Central and South America species of army ant. Sure, there is a certain gory charm in an insect with jaws so large and strong you can use them to hold the sides of a wound together. But it's still difficult to be frightened by something we can squash with a small downward stomp of a hiking-booted foot.

But a colony of army ants, well that's a different matter. A colony probably eats as much flesh as a jaguar does. If you saw a column of them heading your way you might be tempted to put on your boots and run like hell, rather than indulging in a cheery ant-stomping dance.

And so it is with our genome. There are thousands of examples of a particular type of very small junk nucleic acid.[1] Each one plays a role in fine-tuning gene expression, and individually their effects are subtle. But when we look at the totality of their impact, they are an impressive horde.

Welcome to the world of smallRNAs, the mighty army ants of our genome. As their name suggests these RNA molecules are little, typically just 20 to 23 bases in length. We can think of them as nudging molecules, which impart an additional fine-tuning process to control of gene expression.

Figure 18.1 shows how these smallRNAs are produced, and

how they work. They are generated from double-stranded RNA molecules. They then bind to the untranslated regions at the ends of messenger RNAs, to create a new double-stranded RNA. The creation of this double-stranded structure, dependent on the interaction of one junk sequence with another, has one of two effects on the messenger RNA. It can target the messenger RNA for destruction, or it can make it difficult for the ribosomes to translate the messenger RNA sequence into proteins. The end result is essentially similar, a drop in the amount of protein generated from that specific messenger RNA.*[2]

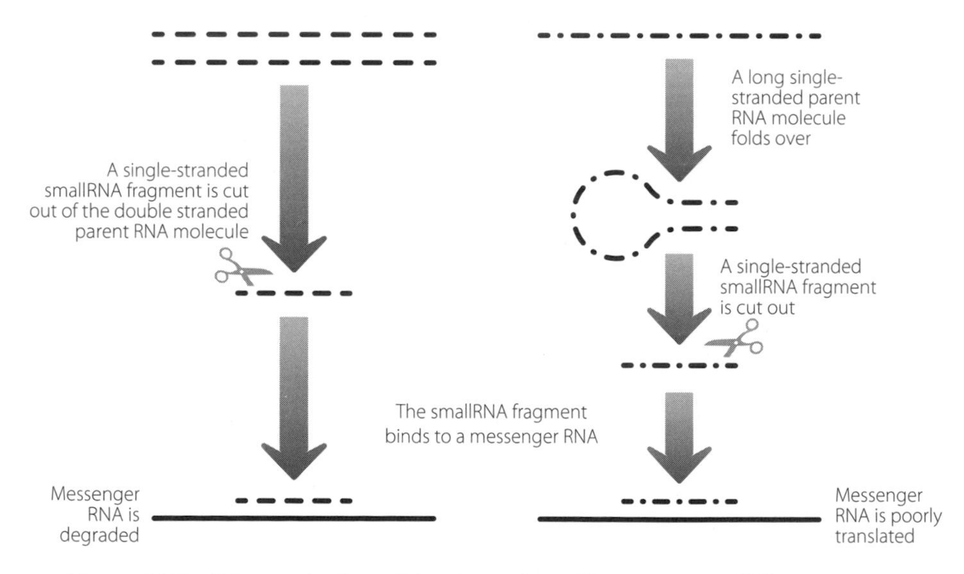

Figure 18.1 Schematic describing how the cell creates two different classes of smallRNAs from longer RNA molecules. The two classes repress gene expression in different ways, as shown at the bottom of the illustration.

* The type of smallRNA that triggers degradation is called microRNA, or miRNA. The type that triggers poor translation is called small interfering RNA, or siRNA. In order to avoid excessive technical language, the term smallRNA will be used to describe both of these.

The smallRNAs that trigger the destruction of messenger RNA molecules have to be a perfect match for their targets. The ones that inhibit the translation of the messenger RNAs are much more promiscuous. They will bind to a messenger RNA even if only a seed sequence of six to eight consecutive bases matches the target. One of the consequences of this is that a single smallRNA may bind to more than one type of messenger RNA, and slow down its translation. Another potential consequence is that the relative levels of the different messenger RNAs in a cell will influence the extent to which each is controlled by a particular smallRNA. This means that any given smallRNA will have a different effect depending on which of its targets is being expressed in a cell, and the ratios of the target molecules.

SmallRNAs – for good, for bad

There is a single cluster of smallRNA molecules that plays an important role in the regulation of a select cell type in the immune system. If this cluster of smallRNAs is over-expressed in mice, the animals develop a fatal over-activation of the immune system.[3,4] On the other hand, mice that lack this cluster altogether die around the time of birth. In humans, the loss of one copy of this cluster leads to some cases of a rare condition called Feingold syndrome.[5] Patients with this disorder have variable symptoms, often including malformations of the skeleton, kidney problems, gut blockages and moderate learning disabilities.[6]

The consequences of disrupted expression of this cluster of just six smallRNAs seem puzzlingly diverse. But perhaps this isn't so surprising, as researchers have calculated that this cluster alone may target over 1,000 protein-coding genes.[7]

The junk sequences that code for smallRNAs are often located within other junk regions, such as the genes producing the long non-coding RNAs.[8] There is a condition called human cartilage-hair hypoplasia, which was originally identified in an Amish community, where one in ten of the community is a carrier of the causative mutation. This is an incredibly high carrier frequency and almost certainly reflects the fact that this community was originally founded by just a small number of families. The affected children have defects in the formation of their skeletons, resulting in a short-limbed form of dwarfism, and light hair that is fine but sparse. The patients also tend to have a variable range of other defects.

The mutations that cause this condition lie in a long non-coding RNA gene. But this long gene encompasses two smallRNA genes, junk within junk, and many of the mutations affect the smaller moieties. The changes disrupt the structures of the smallRNAs so that they aren't processed properly by the cutting enzyme represented by scissors in Figure 18.1. As a consequence, they aren't expressed at their normal level. Between them, these two smallRNAs regulate over 900 protein-coding genes. These include genes known to be involved in skeletal and hair development, but also in a number of other systems. This is presumably why mutations that affect the levels and functions of these smallRNAs can also lead to problems in a range of organ systems in the affected children.[9]

Given how important smallRNAs are for fine-tuning of gene expression, it's perhaps not surprising to learn that these junk molecules have a major role during development. This is the stage in life where apparently minor fluctuations in gene expression can have a significant impact (remember our Slinky falling down the stairs?).

SmallRNAs and stem cells

A beautiful example of the importance of smallRNAs comes from reprogramming human tissue cells to become pluripotent stem cells, potentially capable of building any tissues we need. This is the technology that we first met in Chapter 12, and which is shown in Figure 12.1 (*page 165*). Although the original work for which the Nobel Prize was awarded so quickly was extraordinary, it had some limitations. Although the master regulator proteins could push the developmental Slinky back up a flight of stairs, they did so fairly inefficiently. Only a tiny percentage of cells were converted, and the process took many weeks. Five years after those groundbreaking findings, other researchers extended this work. They treated the adult cells with the same master regulators used in the original experiments. But they also added something else. They over-expressed a cluster of smallRNAs which had been shown to be highly expressed in normal embryonic stem cells. The scientists found that when they over-expressed these smallRNAs along with the original master regulators, adult cells changed back to pluripotent stem cells, as we would expect. But the percentage of cells that converted to stem cells was more than a hundred times greater than with just the master regulators alone. The process also happened much more quickly. Conversely, if they used the master regulators but knocked down the expression of the endogenous smallRNA cluster in the adult cells, the reprogramming efficiency dropped dramatically. This demonstrated that this particular cluster of smallRNAs does indeed play a critical role in helping to regulate the signalling networks that control cell identity.[10,11]

Adult tissues also contain stem cells. These are able to create cells for their specific tissues, rather than multiple cell types. These are important for growth as we move from baby to adult, and also for repairing wear and tear. Some tissues retain a very active stem cell population even late into life. A classic example would be the bone marrow, which keeps producing the cells we need to fight infection and to patrol against potentially cancerous cells. One of the reasons the very elderly are particularly prone to infections and cancer is because their bone marrow stem cells eventually run out, leaving them with holes in their immune barricades.

There are data showing that stem cells and adult cells from human tissues express different patterns of smallRNAs. But expression data are always difficult to interpret, because of the cause-or-effect problem. Are the different patterns of smallRNAs driving the differences in cell activity and function, or are they simply a bystander consequence of the cellular changes? The fact that predicted sequence pairings between individual smallRNAs and the untranslated regions of at least half of all messenger RNA molecules have been preserved through evolution suggests a causal effect.[12] But to address this question more directly, scientists have frequently turned to our close cousin, the mouse.

Researchers have found ways of knocking out genes only in adult tissues, which has created a very powerful tool set for investigations. This handy technique means that mice develop in the usual way, so we don't need to worry that symptoms are caused by pathways and networks going wrong during development. This approach has been used to work out what happens if the enzyme that is required to produce smallRNAs (the scissors in Figure 18.1) is inactivated in adult cells. This will interfere with production of all smallRNAs and so show us where they play an important role. It won't, however, tell us exactly which smallRNAs are involved.

When scientists knocked out the scissors enzyme in all tissues of adult mice, they found defects in the bone marrow, but also in the spleen and the thymus. All three of these tissues produce cells required for fighting infection and were expected to have a large population of stem cells. This finding was consistent with the smallRNA systems having a role in stem cell control. The mice all died, but this was due to a massive deterioration of their intestinal tracts. This is also consistent with a role in stem cells. Our intestines are constantly losing cells that are sloughed off during the continuing activity of the digestive system. These cells have to be replaced every day so we would expect there to be a very active stem cell population.[13] However, it wasn't clear exactly how the loss of the scissors enzyme resulted in dramatic damage to the intestines, although it may have been related to abnormalities in the way the mice processed fats in their diet.

These effects were very dramatic, but that doesn't mean that these are the only tissues where smallRNAs play an important part. Because the mice died relatively quickly, this may have masked more subtle symptoms in other tissues. In order to investigate this, researchers can use a more discriminating version of the adult knockout technique. With this amended technology, they are able to inactivate the scissors gene in selected tissue types in adult mice.

Many of the results were entirely consistent with an impact on stem cell populations. For example, when the scissors gene was inactivated in the cells of the hair follicle in adult mice, fur didn't grow back properly after plucking.[14]

It would be tempting to speculate from these results that the smallRNA networks are required to keep stem cells doing their job of replenishing specialised cells. But this is too simplistic. Just as we all strive to make our salary last until the next payday, our bodies need to make sure they don't use up their stem cells too quickly. They are precious, and when they're gone, they're gone. Once we appreciate that, it seems obvious that some smallRNA networks are required to stop stem cells from irreversibly converting into mature tissue cells. There is actually a balance that needs to be struck, and this is shown in Figure 18.2.

The skeletal muscles contain stem cells,* and it's worth keeping these quiescent most of the time so that they don't get used up too early. This exhaustion of the stem cell reservoir is partly responsible for some of the muscle loss we have encountered already in conditions such as Duchenne muscular dystrophy. There are proteins in muscle stem cells that normally stop them from converting into mature muscle cells. However, if there is an acute injury in

* These are known as satellite cells

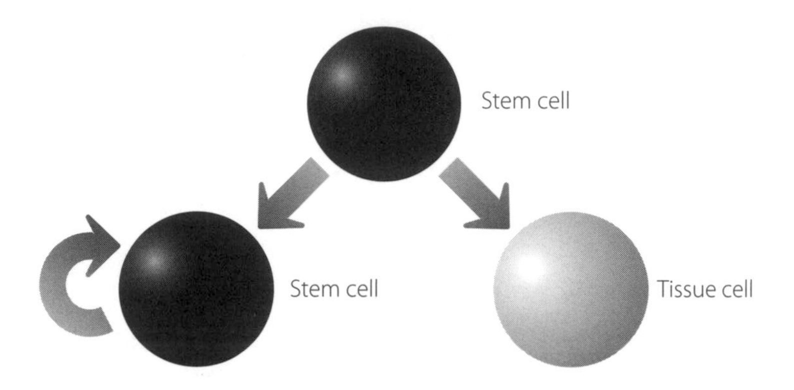

Figure 18.2 When a stem cell divides it can create either another stem cell, which can also keep dividing, or a differentiated cell that will not create more stem cells.

healthy individuals, or loss of muscle cells in a dystrophic condition, these proteins are down-regulated. This is achieved at least in part by switching on expression of specific smallRNAs. The smallRNAs bind to the messenger RNAs that carry the code for these proteins, and less protein is produced. The brakes are taken off the stem cells and they convert into mature muscle.[15,16]

A similar effect can be seen in the heart. The adult cardiac muscle does contain some stem cells, although they aren't huge in number and they are hard to convert into mature heart tissue. This is one of the reasons why heart attacks are so damaging. In a heart attack, cardiac muscle dies and our bodies find it very difficult to create replacement tissue. Instead, we get scarring on the heart and the organ doesn't work properly. This leads to the long-term difficulties many heart attack survivors encounter, and is why in some cases they never regain full health.

Although it might seem that it would be great to be able to activate cardiac stem cells to produce new muscle, experiments from mice suggest that the situation isn't straightforward. It would seem that in the heart the smallRNAs prevent stem cells converting into cardiac muscle. If the scissors enzyme that produces smallRNAs is switched off in an adult heart, the heart begins to grow. Unfortunately, it does so in a way that is potentially damaging, resulting in a condition known as cardiac hypertrophy. This is unlike the helpfully strong heart muscle of elite athletes. Instead it is more like the abnormal thickening of the heart walls found in people with high blood pressure. Its seems to happen because loss of scissors activity causes the stem cells to stop acting like adult cells and drives a gene expression pattern that's more like the one seen during development.[17]

It might seem odd that reactivating cardiac stem cells isn't necessarily helpful but perhaps it's a trade-off. In evolutionary terms, the most important consideration for animals is to live long enough to reproduce and pass on one's genetic material. The control of cardiac development is geared towards making sure that our hearts are good enough to get us to this point. From an evolutionary perspective, it doesn't really matter if this means that when we are older we can't repair our hearts. This is a problem for humans because we like living longer than evolution deems strictly necessary.

SmallRNAs and the brain

Although we usually think of our brains as being fully formed in adults, recent data have shown that even in this organ there are some stem cells. In animals that rely on a highly developed sense of smell, these stem cells can be activated to form neurons that respond to new scents. This allows the animal to tailor the smells to which it responds most strongly. A protein in the stem cells drives them into differentiating into a specific type of responsive neuron. Expression of this protein is usually held in check by a smallRNA. When researchers inhibited the expression of this smallRNA in mice, the protein was up-regulated and the neural stem cells differentiated into neurons associated with detection of smell.[18] The suspicion is that the smallRNA is down-regulated naturally when the mouse smells something new, although the signalling pathways that drive this repression haven't been identified yet.

SmallRNAs are involved in everyday cellular activities, fine-tuning responses to constantly fluctuating environments. It can be difficult to unravel how this fine-tuning operates, because each individual smallRNA has a relatively small effect. It's the overall cumulative effect of multiple smallRNAs acting in vast but subtle networks that is their most important feature. Even so, enough intriguing data are emerging to give us confidence that this class of miniature junk minions has real impact.

The brain appears particularly sensitive to perturbations of the smallRNA landscape. The impact of such changes varies depending on the regions of the brain involved, but also on the timing of the perturbations. This in turn probably reflects the importance of cross-talk between all the different smallRNAs and all the other messenger RNAs and proteins whose expression is tightly controlled in the brain.

A striking example of this is found when the scissors enzyme is inactivated in a region called the forebrain in adult mice.[19] The expression of smallRNAs is lost, and at first it seems like this is quite a good thing for the animals. For about three months the mice are smarter than usual. They perform better at tests, whether these are based on fear or on reward. Their memory skills are significantly improved. But in case anyone is thinking of trying this at home on their own brain (everyone is very exam-focused these days), there is a downside. The intellectual star of these smart mice shone brightly, but it didn't shine for long. About twelve weeks after the scissors enzyme was inactivated, the brains of the furry little geeks began to degenerate.

This delayed reaction was also found in another situation where smallRNAs were shown to be important in the brain. This may imply that smallRNAs are fairly stable in brain cells, and take a while to die down. The scissors enzyme was inactivated in brain cells of two-week-old mice, in a region that is involved in the control of movement. As expected, this resulted in a major drop in the expression of smallRNAs. The mice appeared fine at first but eleven weeks later they began to develop movement problems. Analyses of their brains showed that the neurons that lacked the ability to make smallRNAs had died.[20]

SmallRNAs can turn up in all sorts of unexpected situations. One of the targets for alcohol in our brains is a protein that regulates how signals pass across the membranes of cells.* The messenger RNA for this protein can occur in lots of different versions, depending on how the amino acid-coding regions are spliced together. Alcohol induces the expression of a particular smallRNA which can bind to the untranslated region at the end of some of these variant messenger RNAs. This leads to selective destruction of the messenger RNAs that code for some variants of the proteins, but not others. This change in the population of the possible proteins leads to a skewing in the responses of the neurons to alcohol, and is an important part of the tolerance to alcohol that is a component of addiction.[21] This mechanism is summarised in Figure 18.3. SmallRNAs have also been implicated in addictive responses to other drugs, such as cocaine.[22]

Small RNAs and cancer

Mis-expression of smallRNAs has been implicated in a number of diseases that have a major impact on global human health. These include cardiovascular diseases[23] and cancer.[24] The latter is perhaps unsurprising, given that cancer represents abnormalities in cell fate and cell development, and smallRNAs are very important in these processes. One very clear example of the importance of smallRNAs in cancer is in a type of tumour that is characterised

---

* This is a protein called BK, which is a potassium channel.

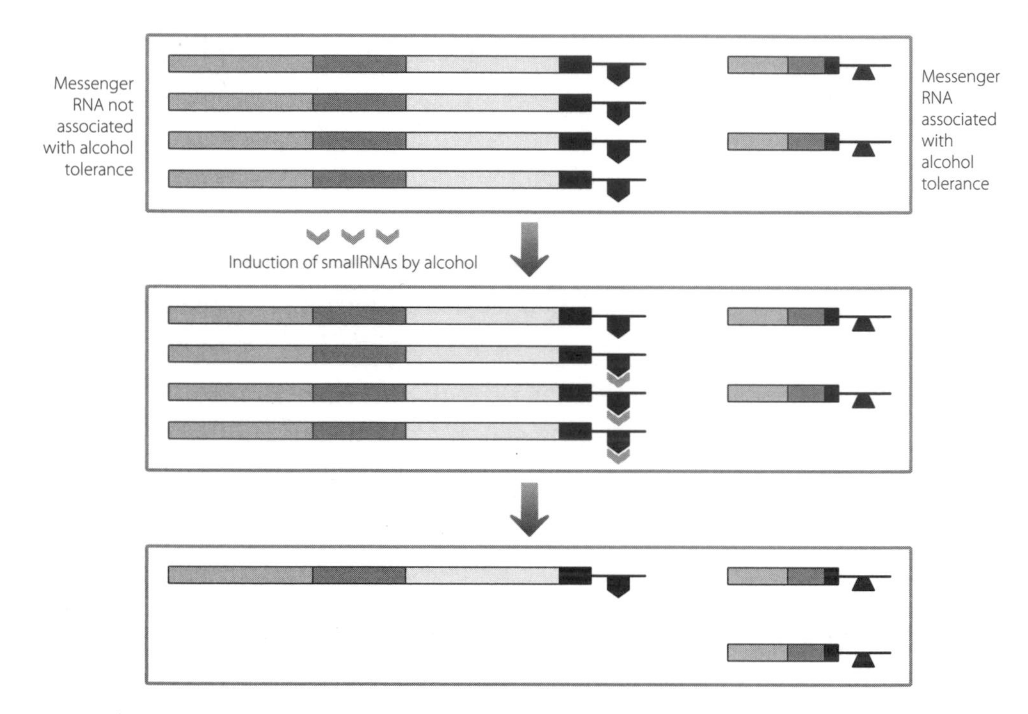

Figure 18.3 SmallRNAs induced by alcohol can bind to messenger RNAs that don't create alcohol tolerance. The smallRNAs don't bind to the messenger RNA molecules that promote alcohol tolerance. This leads to a relative preponderance of the messenger RNA molecules that code for protein versions associated with tolerance to alcohol.

by inappropriately expressing developmental rather than postnatal genes. It's a subtype of a childhood brain tumour which usually presents before the age of two. Sadly, it's a very aggressive form of cancer, and the prognosis is poor even with powerful therapy.* The cancer develops following an inappropriate rearrangement of genetic material in the brain cells. A promoter that normally drives strong expression of a protein-coding gene recombines with a particular smallRNA cluster. This whole rearranged region is then amplified, meaning multiple copies are produced in the genome. As a consequence, the smallRNAs downstream of the

* These are known as supratentorial neuroectodermal tumours.

relocated promoter are expressed far too strongly. The levels of the smallRNAs are between 150 and 1,000 times higher than they should be.

The cluster codes for over 40 different smallRNAs, and is in fact the largest cluster in primates. It is usually only expressed early in human development, in the first eight weeks of foetal life. Switching it on strongly in the brain of an infant has a catastrophic effect on gene expression. One of the downstream effects of this is to drive expression of an epigenetic protein which adds modifications to DNA. This leads to global changes in DNA methylation patterns, resulting in abnormal expression of a whole range of genes, many of which should be expressed only when the immature brain cells are dividing during development. This generates a cancerous cell programme in the infant.[25]

This cross-talk between smallRNAs and the epigenetic machinery of the cell may be significant in other situations where cells become predisposed to cancer. This mechanism can amplify the impact of disrupted smallRNA expression, by altering epigenetic modifications, which can be passed on to daughter cells. This can start a hard-wiring in of potentially dangerous alterations in gene expression.

Not all the steps have been unravelled in how smallRNAs interact with epigenetic processes, but hints are emerging. For example, a particular class of smallRNAs which trigger increased aggressiveness in breast cancer targets the messenger RNAs for certain enzymes that remove key epigenetic modifications. This alters the pattern of epigenetic modifications in the cancer cell, and further disrupts gene expression.[26]

Many cancers are surprisingly difficult to monitor in a patient. They may be inaccessible, so that they are hard to sample. This can make it difficult for clinicians to monitor how a cancer is changing, and exactly how it is responding to therapies. They may have to rely on indirect measures, such as imaging the tumour on a scan.

Some researchers have suggested that smallRNA molecules may provide a new technique for following the natural history of a tumour. When cancer cells die, this often results in the smallRNAs leaving the cell as it breaks down. These little junk molecules are often complexed with cellular proteins, or wrapped in fragments of the cell's membranes. This makes them very stable in body fluids, so they can be isolated and analysed. Because the amounts are low, researchers need to use very sensitive analytical techniques. This isn't impossible though, because nucleic acid sequencing sensitivity is improving all the time.[27] Data in support of this approach have been published for breast[28] and ovarian cancer,[29] among others. In the case of lung cancer, circulating smallRNAs have been analysed and shown to be useful at discriminating between patients with a solitary lung nodule that is benign (doesn't require therapy) from patients where the nodule is a tumour (and needs treatment).[30]

Dead horses and silenced genes

SmallRNAS are turning up in all sorts of unexpected situations. There is a really horrible viral infection called North American eastern equine encephalitis virus. It's transmitted by mosquito bites. When this virus infects horses, the animals die. The situation isn't much better in humans, where the fatality rate is between 30 and 70 per cent. The patients die because the virus gets into the central nervous system and causes severe inflammation of the membranes around the brain.[31] The virus that causes the infection has a genome that is made of RNA, not DNA.

When this virus first enters the human bloodstream following a mosquito bite, it is taken up by white blood cells. These are the front line in surveillance against invaders. But then something very odd happens. A smallRNA naturally produced by the white blood cells binds to the end of the virus's RNA genome, and stops it from coding for protein.

This might seem like a good thing but it's quite the opposite. Our white blood cells normally recognise if they have been infected by a virus. The cells will initiate a set of reactions including raising body temperature, and producing various anti-viral chemicals. Together, these repel the tiny invaders.

But when the smallRNA in the white blood cells binds to the equine encephalitis virus genome, the virus goes quiet. Consequently, the immune system doesn't notice that the body has been infiltrated. This leaves other viral particles free to drift through the body. If some of them reach the central nervous system, they can then trigger the lethal responses in the brain tissues.[32]

The researchers described this in terms of the virus hijacking the smallRNA system, and it doesn't seem to be the only example of this happening. The hepatitis C virus also has an RNA genome. When this virus infects liver cells, the viral RNA binds to a smallRNA naturally expressed by these cells. In this case, the binding stabilises the viral genome, making it harder to break down. As a consequence, more viral proteins are produced, and the infection becomes more damaging and more aggressive.[33]

It's pretty clear that smallRNAs are involved in a whole range of human pathologies from infection to cancer, and from development to neurodegeneration. This of course raises an interesting question: if junk DNA can cause or contribute to disease, is it also possible to use junk to fight common human illnesses?

19. The Drugs Do Work (Sometimes)

Billions of dollars are spent every year by companies trying to create new drugs to treat human diseases. They hope to find ways to tackle unmet medical needs, a situation that is becoming ever more urgent with the increasing age profile of the global population. The breakthroughs in the understanding of the impact of junk DNA on gene expression and disease progression are triggering a slew of new companies seeking to exploit this field. Specifically, most of the new efforts are in using non-protein-coding RNAs as drugs in themselves. The basic premise is that junk RNA – long non-coding, smallRNAs or another form called antisense – will be given to patients, to influence gene expression and control or cure disease.

This is very different from the way we treat diseases at the moment. Historically, most drugs have been of a type known as small molecules. These are chemically created and are relatively simple in shape. Examples of some common small molecule drugs are shown in Figure 19.1.

More recently, we have learnt how to use proteins as drugs. Probably the most famous is insulin, the hormone that diabetics use to regulate their blood sugar levels. Antibodies are another very successful type of protein drug. These are engineered versions of the molecules we all produce to fight infections. Drug companies have found ways of adapting these so that they will bind to over-expressed proteins and neutralise their activities. The bestselling antibody is one that treats rheumatoid arthritis very effectively, but

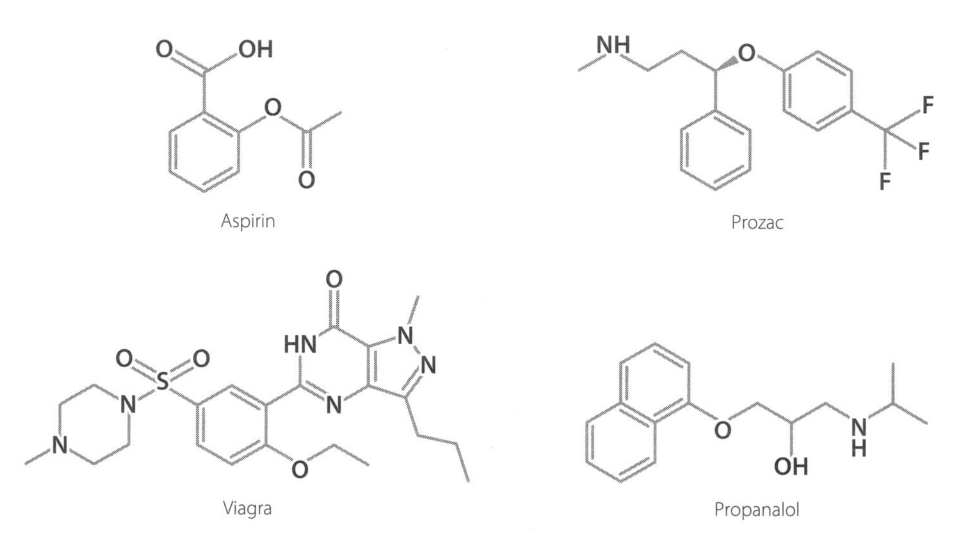

Figure 19.1 Structures of some commonly used small molecule drugs.

there are others that treat conditions as diverse as breast cancer and blindness.[1]

Small molecules and antibodies have advantages and drawbacks. Small molecules are usually relatively cheap to synthesise and easy to administer, frequently just needing to be swallowed. Their drawback is that they don't last very long in the body, which is why we need to take them on a regular basis. Antibodies can last for weeks or even months in the body, but they have to be injected by a medical professional, and they are very expensive to manufacture.

There are some other drawbacks too. Antibodies are only effective against molecules that are in body fluids such as blood, or are on the surface of cells. These drugs can't get inside cells to do their work. Depending on their structure, small molecules can get inside cells if necessary. But they may be limited in the kinds of proteins that they can control.

Small molecules act like a key in a lock. If you are inside your house, the easiest way to stop someone coming in is to lock your door and leave the key in it. If you wanted to stop anyone else from ever entering, you could even lock the door using a slightly defective key, which jams the lock for ever.

This works because the key fits into the lock really snugly. But what you can't do is use a key to block one of those old-fashioned external sliding bolts. There is nowhere for the key to fit in this, it will just keep slipping around on the surface. This is also true of our cells. There are lots of proteins inside our cells that we would like to control but we can't create small molecules against them, because of the protein structure. They don't have nice neat clefts or pockets that we can fit drugs into. Instead, they have large flat surfaces, and there is nowhere for a small molecule to lodge.

We could try to make bigger molecules that can cover the whole flat surface. The problem with this is that once we get above a certain size with our drugs they don't circulate well around the body, and they can't get into the cells to do their job.

There's also another problem. It's hard enough to create drugs that will successfully get inside cells, bind to a specific protein, and stop that protein from working. But it's incredibly difficult to create drugs that will get inside cells, bind to a specific protein and then make that protein work harder, or faster, or better. And it's practically impossible to make traditional drugs that will drive up the expression of one specific protein, or switch on one and only one gene.

Could junk DNA save us?

This is why there is so much interest in finding new approaches to drug therapies, and why the increasing knowledge of junk DNA is so important. By using long non-coding RNAs or smallRNAs, it is theoretically possible to target pathways that can't be tackled using traditional small molecule or antibody drugs. It won't matter that the targets are inside cells and have large flat surfaces. It won't matter that we need to increase expression or activity of a protein or gene. We can use this new approach to tackle any type of target.

Theoretically.

That's the word to focus on. Theoretically. Ideas are common, success is rare. So it's worth taking a good look at where the reality is before we all cash in our pensions to invest in the latest biotech company working in this space. There is a lot of activity going on,[2] so it's worth concentrating our analysis on a few leading examples.

There is a protein produced in the liver that is responsible for transporting some other molecules around the body. Globally, there are about 50,000 people who have inherited a mutation in the gene that produces this protein. There are lots of different mutations, but they all seem to have a similar effect. They all change the activity of the protein so that it starts to transport the wrong molecules.*[3]

When this happens, deposits, which include a mixture of normal and mutated protein, begin to build up in tissues. Patients may have a range of symptoms, depending on the tissues in which the deposits build up. In about 80 per cent of known cases, the heart is the main organ that is affected, and this leads to potentially lethal cardiac defects. In many of the other 20 per cent of cases, the deposits build up in the nerves and spinal cord. This can lead to debilitating problems with a range of organs, including abnormal and painful sensory responses to mild stimuli.

A company called Alnylam has created a smallRNA, attached to sugar molecules, which can be injected into patients. The smallRNA binds to the untranslated region at the end of the messenger RNA that codes for the protein that is mutated in this disease. This targets the messenger RNA for destruction.

In 2013 the company released data from a phase II clinical study of their drug. They found that when they injected the drug

* The protein is called transthyretin.

into patients, there was a rapid and sustained drop in the circulating levels of the mutated and normal versions of the protein.[4] This is encouraging, but not yet a cure. The assumption is that a drop in circulating levels will lead to a slowdown in the build-up of tissue deposits. This in turn should lead to at least a slowdown in the progress of the disease. But we won't know if that is the case until a bigger trial is carried out, in which the actual symptoms and disease progression are monitored. Only if the drug impacts on these will it be considered a success.

A different company, called Mirna Therapeutics, has created a smallRNA which mimics one known to be important in cancer. The endogenous smallRNA is a tumour suppressor, and its overall effect is to hold back cell proliferation. It does this by negatively regulating the expression of at least twenty other genes that try to push the cells into division. Expression of this smallRNA is often lost or decreased in cancer patients, removing the brakes on cell division. The hope is that by reintroducing it into cells, the normal pattern of gene regulation will be restored, and the cells will stop proliferating so quickly.

The company has tested their mimic in patients with liver cancer. So far the trials have just been designed to see what doses of the drug the patients are able to tolerate. It will be some time before we will know if this approach is going to result in clinical benefit.[5]

There is a clever, although not immediately obvious, angle to the products being developed by both Alnylam and Mirna. One of the biggest problems that companies have faced in the past when trying to develop drugs around nucleic acids has been the body's own detoxification abilities. This is also often a problem for traditional drug discovery as well. Essentially, when a new chemical of any type enters the body, there is a very high likelihood that it will go to the liver. One of the main jobs of this vastly energetic organ is to detoxify anything it doesn't like the look of. For all of our evolutionary history, this has served us well, protecting us from toxins in food. But the problem is that the liver has no means of distinguishing between toxins we want to avoid, and drugs we are trying to use. It will just drag them in, and try to destroy them.

To use an old rubric, Alnylam and Mirna are making a virtue from a necessity. Alnylam is targeting expression of a protein that is produced in the liver. Mirna is trying to develop treatments for liver cancer. Their molecules will be taken up by exactly the organ they want them to reach. The companies have adapted the structure or packaging of their molecules to try to ensure that once they are in the liver, they will survive long enough in the cells to do their job. SmallRNA approaches have been put forward for a number of other conditions, and the preliminary cellular and animal experiments often look good. But for a condition such as amyotrophic lateral sclerosis, where the nucleic acids will have to avoid the liver and be taken up by the brain,[6] it's not clear yet how successful the industry will be in capitalising on this technology.

In Chapter 17, we saw how hopes of a promising new approach to treat Duchenne muscular dystrophy may be receding, after unexpectedly disappointing late-stage clinical trial failure. The methodology used in this approach was an example of a particular kind of junk DNA, known as antisense.

Antisense junk RNAs are probably a widespread feature of our genome, and it's because of the double-stranded nature of DNA. We touched on this in Chapter 7, where the actual biological example we used was of Xist and its antisense counterpart, Tsix. We also used the analogy of the word DEER, which can be read backwards as the word REED. It just depends if the enzymes that make RNA copies of DNA reads one strand from left to right, or the opposite strand from right to left.

However, most words can't be read in both directions. If we read the word BIOLOGY backwards, we get YGOLOIB, which doesn't have a meaning. In the same way, messenger RNA from one direction in the genome may code for a protein, but the same region copied backwards simply codes for a junk RNA that cannot be translated into a protein. Sometimes this creates auto-regulatory loops in our cells, limiting expression of certain genes. An example of this is shown in Figure 19.2.

Researchers have reported that about a third of protein-coding genes also produce junk RNA from the antisense strand. However, the antisense is usually produced at lower levels, often no more than 10 per cent.[7] Sometimes the antisense is just a short internal section of the gene. Other times the sense and antisense may start and end in different places so that they overlap but also have unique regions. Sometimes the machinery copying the sense DNA strand into sense RNA crashes into the machinery moving in the other direction to create the antisense RNA. Both sets of proteins fall

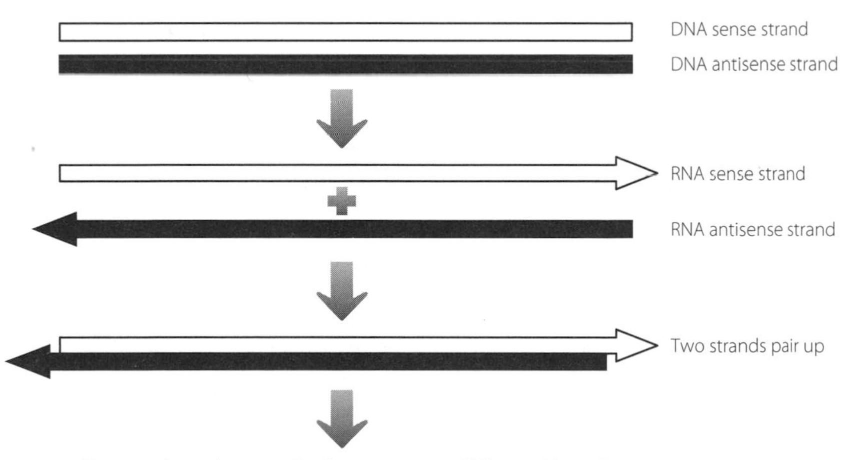

Figure 19.2 In some parts of the genome, both strands of DNA can be copied into RNA, in opposite directions. These are known as sense (creating RNAs that code for protein sequences) and antisense (which don't code for protein sequences). The antisense RNA molecule can bind to the sense RNA molecule and affect its activity, in this example by inhibiting production of protein from the sense messenger RNA template.

off the DNA, and both RNA molecules are abandoned. There are even antisense strands for some long non-coding RNAs.

The effects of an antisense RNA binding to its sense RNA partner can vary. Figure 19.2 shows an example where this binding prevents the sense messenger RNA from being translated into protein. But there are other situations where the binding stabilises the messenger RNA, ultimately leading to higher protein expression.[8]

In the Duchenne muscular dystrophy trials that originally held such promise, the patients were treated with an antisense molecule that could recognise and bind to messenger RNA for dystrophin. The antisense molecule was chemically modified to prevent it from being broken down too quickly in the body. When the antisense molecule bound to the dystrophin messenger RNA, it prevented the splicing machinery from binding in the normal way. This altered the way the messenger RNA was spliced together, and got rid of the region that caused the most problems in mutant protein production.

There are some happy endings

The Duchenne muscular dystrophy trial ultimately failed but we shouldn't take this as meaning the entire antisense field is tainted. In fact, it's had its successes. In 1998 an antisense drug was licensed for use in immunocompromised patients who had developed a viral infection in the retina* that threatened their sight. The antisense molecule bound to a viral gene, and prevented the virus from reproducing.[9] It was an effective drug, which raises two questions. Why did this drug work so well? And given that it worked so well, why did the manufacturer stop selling it in 2004?

Both answers are quite straightforward. The drug worked well because it was injected straight into the eye. There was never a

* The virus was cytomegalovirus (CMV).

problem about it being scooped up by the liver, because it didn't go via the liver. It was also targeting a virus, and only in one self-contained part of the body, so there wasn't much risk of widespread interference with human genes.

All of which sounds peachy, so why did the manufacturer stop selling it in 2004? This drug was developed for severely immunocompromised patients, of whom the vast majority were people suffering from AIDS. By 2004, there were drugs available that were pretty good at keeping HIV, the causative virus, under control. The patients' immune systems were in much better shape, and they simply weren't succumbing to viral infections in the retina anymore.

More recent developments have also shown that there is still life in the use of antisense junk DNA for therapy. There is a serious condition called familial hypercholesterolaemia. In the UK it is predicted that there are about 120,000 people with this disorder, although many of them may not have been diagnosed. These people have genetic mutations that prevent their cells from taking up bad cholesterol and dealing with it properly. As a consequence, between a third and half of all such patients will have serious coronary artery disease by their mid-50s.[10]

For some patients with this condition the standard lipid-lowering drugs, known as statins, work really well to lower their risk of cardiovascular disease. This is often the case for people who have one mutant copy of a particular gene, but in whom the other matching copy is normal. But there are some severely affected patients, especially those in whom both copies of the specific gene are mutated, for whom statins are ineffective. These patients often have to undergo plasmapheresis once or twice a week, where their blood is passed through a machine and the dangerous cholesterol is removed.

If you want to stop a bathtub from overflowing, you have two options. You can keep letting water out via the drain, or you can turn down the taps to stop adding more water.

A company called Isis developed an antisense molecule which targets the primary protein in low-density lipoproteins, the so-called 'bad cholesterol'.* This antisense therapy for familial hypercholesterolaemia works by turning off the taps. The antisense drug binds to the messenger RNA for the bad cholesterol protein and suppresses it, resulting in lower expression and lower levels of bad cholesterol. Isis licensed this to a larger company called Genzyme, in a deal costing hundreds of millions of dollars.

This antisense drug** was licensed for use by the US Food and Drug Administration in January 2013. It is only licensed for patients who suffer from the most severe form of familial hypercholesterolaemia. One of the reasons this drug has been successful enough to reach the market (albeit at the eye-watering cost of over $170,000 per year per patient[11]) is because the gene that it targets is expressed in – yes, you guessed it – the liver. A downside to this, however, is that there have been liver toxicities reported with the use of this drug. The Food and Drug Administration has demanded that Sanofi (who bought Genzyme) must monitor liver function of all patients.[12] The European Medicines Agency refused to license the drug at all, citing safety concerns.[13]

The hundreds of millions of dollars that Isis received from Genzyme for its antisense therapy is a lot of money. Yet consider this. It took over twenty years to move from the basic research to a marketed drug, and the whole process cost over $3 billion.[14] That's an awfully big investment to recoup.

Of course, pioneering drugs, especially those which use a relatively untried type of molecule, would be expected to take a long time and a lot of money to develop. The hope is always that later programmes are able to run faster and more smoothly. Certainly, clinical trials for therapies based around junk DNA are building

* The protein that is targeted is called apolipoprotein B100.

** The drug is called Mipomersen, also known as Kynamro.

in number. There is a human smallRNA that is co-opted by a virus to help it infect cells. In an example of using junk to fight junk, an antisense drug is in phase II clinical trials, targeting this smallRNA.[15]

But here's an odd thing to consider. In 2006 the pharmaceutical giant Merck paid over a billion dollars for a company that was developing smallRNAs as therapeutics. In 2014 it sold the company on, for a fraction of what it paid.[16] Another company, Roche, stopped its own efforts in this research area in 2010.

There has recently been a big upsurge in investment into biotech companies working on smallRNAs. RaNA Therapeutics, which is believed to be developing RNA-based drugs that will prevent the interaction of long non-coding RNAs with the epigenetic machinery, raised over $20 million in 2012.[17] Dicerna, which is developing smallRNAs against some rare diseases and oncology indications, raised $90 million in 2014.[18] That's the third set of financing it has received, despite having no programmes that have reached clinical trials yet.[19]

Yet here's the weird thing. Literally as I write this chapter, in spring 2014, an alert comes up in my email account and tells me that Novartis has decided to slow down dramatically its research on this topic.[20] The pharma giant mainly cited the ongoing problems with working out how to deliver smallRNAs to the right tissues. This has been the biggest issue with these therapeutics since companies first started trying to develop them. Many of the companies in the junk RNA field have been set up by brilliant scientists, but that doesn't mean that any of the basic drug delivery problems will just disappear overnight. Not all the companies will fail. But quite a lot of them probably will. There haven't been any major breakthroughs on this problem, and certainly nothing that would explain why investors are pouring money into new biotechs in this area.

One day science will probably be able to interpret all the possible epigenetic modifications that are found in the genome and predict precisely what their consequences will be for gene expression. We'll work out how to capture carbon, and how to establish colonies on Mars. Tuberculosis will be a distant memory and we'll all have a good grasp of the Higgs boson. But unravelling the reasons behind the triumph of hope over experience in the investment community? Be realistic.

20. Some Light in the Darkness

As we near the end of our wanderings through the darker regions of our genome, the more alert reader may remember that we haven't addressed the mystery of one of the human disorders first encountered at the beginning of this book. This condition is the cumbersomely named facioscapulohumeral muscular dystrophy, or FSHD. This is the condition in which there is wasting of the muscles of the face, shoulders and upper arms.

It occurs when patients inherit a smaller number of a particular genetic repeat on one of their copies of chromosome 4. Even quite a few years after the mutation was identified, the reason why this caused disease remained mysterious, because there just didn't seem to be a protein-coding gene anywhere near the genetic defect.

We finally have an understanding of how the disease symptoms are caused and the story is remarkable. It pulls together a number of the themes we have already encountered, showing how junk DNA, epigenetics, genetic fossils and abnormal RNA processing all work together to create an extraordinary tale of pathological conspiracy.[1]

Let's recap a little. On normal copies of chromosome 4, a region is repeated between eleven and 100 times. This region is just over 3,000 base pairs in length. In people with FSHD, there is a much smaller number of repeats – between one and ten units – on one of their copies of the chromosome.

Here's where the first complication arises. There are people who have ten or fewer copies of this unit, but who don't have

FSHD. Their muscles are completely healthy. The low number of repeats only causes a problem if it occurs on a copy of chromosome 4 that also contains another feature.

To understand the importance of this other feature, we need to look in more detail at what is found in the repeating units. They all contain a retrogene.* A retrogene is a form of junk DNA. It is created when the messenger RNA from a normal cellular gene gets copied back into DNA and reinserted into the genome. It's very similar to the process we saw in Figure 4.1 (*page 38*) and occurred long ago in human evolution.

Because retrogenes are originally created from messenger RNA templates, they often don't include the proper regulatory sequences of normal genes. They won't contain splicing signals (because the messenger RNA template had already been spliced before it was copied into DNA) and they lack appropriate promoter and enhancer regions. But some can still be used to produce messenger RNA. This is the case with the FSHD retrogene, but it doesn't usually matter, because the RNA doesn't function properly in the cell. It doesn't contain the correct signals for adding a string of A bases to the end of the messenger RNA, the process described in Figure 16.5 (*page 233*). Because of this, the messenger RNA is unstable and doesn't get used as a template for production of protein.

But, when a person has only a small number of FSHD repeats, and other sequences on chromosome 4 are present, the final copy of the FSHD retrogene can be spliced to an additional sequence. This creates a signal at the end of the messenger RNA which allows the cellular machinery to add A bases. This in turn stabilises the messenger RNA, and it is transported to the ribosomes to act as the template for production of a protein – a protein that should never be switched on in mature muscle cells.

---

* This particular retrogene is called DUX4.

The FSHD protein is one that regulates the expression of other genes by binding to specific DNA sequences. It is usually only expressed in the germline, the cells that produce eggs or sperm. There is no definitive explanation yet for why expression of this protein causes muscle wasting, and it may be that a number of mechanisms are involved. It may activate genes that trigger muscle cell death. It may cause loss of muscle stem cells, perhaps by activating other retrogenes and genomic invaders that should be kept silent. One intriguing possibility is that the muscle cells that express the FSHD protein are destroyed by the patient's own immune system.

The germline is a tissue that is known as immunologically privileged, because normally it is kept isolated from the cells of our immune system. This means that our immune system never learns that the cells of immunologically privileged sites are a normal part of our body. If proteins from the germline are switched on in adult muscle cells, the immune system may respond as if they are foreign organisms and attack the cells that express these previously unencountered elements.

So FSHD provides us with an example of the importance of junk DNA in disease. A genetic defect changes the amount of junk DNA. As a consequence of this, a junk element is expressed and modified by the addition of a junk sequence. But there is yet more to the picture. The FSHD retrogene only becomes stably expressed in the presence of a particular pattern of epigenetic modifications.

In normal cells, the FSHD repeats are usually expressed when the cells are in a pluripotent state, such as embryonic stem cells. At this stage, the FSHD repeats are covered with activating epigenetic modifications. But as the cells differentiate, the activating modifications are replaced by repressive ones, and the region is silenced. But if pluripotent cells are created from FSHD patients, the activating modifications aren't replaced as the cells differentiate, and the repeats remain switched on.

Another aspect of the picture is the overall control of the FSHD genetic domain. There is an insulator region between the repeated region and the rest of chromosome 4. The protein 11-FINGERS (*see page 178*) binds to this region and ensures that different patterns of epigenetic modifications are maintained in the FSHD domain compared with the adjacent areas of the chromosome.

On top of all these features, the three-dimensional structure of the relevant regions of chromosome 4 also plays a role in the expression of the FSHD retrogene. It's almost certainly the combination of all these factors that results in the restricted pattern of muscle wasting that we see in patients with FSHD. All of these aspects have to be right (or perhaps wrong) for the symptoms to develop.

The mechanism by which a change in a junk region leads to disease in FSHD is a stunning example of the complex and multi-layered ways in which the different elements of our genome work together. It also demonstrates how we need to think not in terms of linear pathways when we consider what is happening in our cells, but in terms of complex interlocking processes. Figure 20.1 demonstrates this graphically. It exemplifies why the arguments about which is the most important feature of our genome are ultimately sterile. If we disrupt any aspect there will be consequences. Some will be bigger than others, but all work together.

Of course, this doesn't mean that every single one of our billions of base pairs has a function. Some may truly just be genomic garbage, with no utility, whereas other regions are junk in the sense that they could have been discarded but instead have been turned into something useful.[2]

There is still a lot we don't know, including some questions that we might think are very straightforward. We haven't even got a definitive answer for how many functional regions of junk DNA exist in a cell. That might seem easy to answer but have a quick

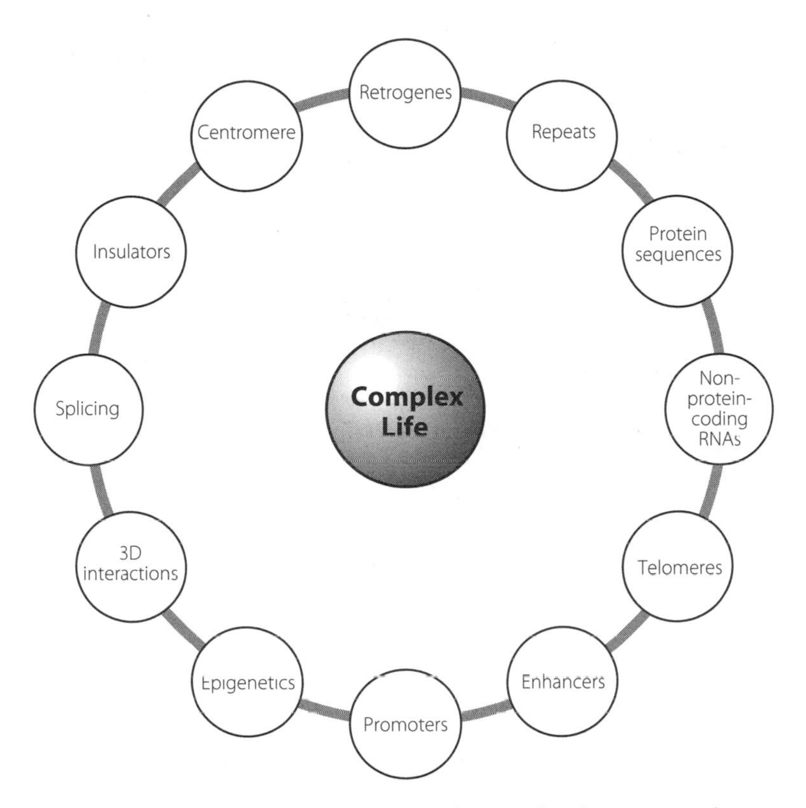

Figure 20.1 Just some of the interacting factors that have to work together to create the great organism that is you.

look at Figure 20.2 and then answer the following question. How many squares are there on a chessboard?

The instant instinctive response is always 64. But the actual answer is 204, because we can draw bigger squares of various sizes around the more obvious single black and white ones. Our genome is like that. One stretch of DNA can include a protein-coding gene, long non-coding RNAs, smallRNAs, antisense RNAs, splice signal sites, untranslated regions, promoters and enhancers. Layer on to this the effects of variations in DNA sequence between individuals, directed and random epigenetic modifications, changeable three-dimensional interactions, plus binding to other RNAs and proteins; then add in the effects of our constantly altering environment.

Figure 20.2 Quickly now, how many squares are on a chessboard?

When we really think about the complexity of our genomes, it isn't surprising that we can't understand everything yet. The astonishing triumph is that we understand any of it. There is always something new to be learnt, out there in the dark.

Notes

Chapter 1

1. For information on the disorder and its genetics see www.omim.org record #160900
2. For more information see http://ghr.nlm.nih.gov/condition/myotonic-dystrophy
3. For more information see http://www.ninds.nih.gov/disorders/friedreichs_ataxia/detail_friedreichs_ataxia.htm
4. For more information see http://ghr.nlm.nih.gov/condition/facioscapulohumeral-muscular-dystrophy

Chapter 2

1. http://www.escapistmagazine.com/news/view/113307-Virtual-Typewriter-Monkeys-Pen-Complete-Works-of-Shakespeare-Almost
2. Campuzano V, Montermini L, Moltò MD, Pianese L, Cossée M, Cavalcanti F, Monros E, Rodius F, Duclos F, Monticelli A, Zara F, Cañizares J, Koutnikova H, Bidichandani SI, Gellera C, Brice A, Trouillas P, De Michele G, Filla A, De Frutos R, Palau F, Patel PI, Di Donato S, Mandel JL, Cocozza S, Koenig M, Pandolfo M. Friedreich's ataxia: autosomal recessive disease caused by an intronic GAA triplet repeat expansion. *Science*. 1996 Mar 8;271(5254):1423–7
3. Bidichandani SI, Ashizawa T, Patel PI. The GAA triplet-repeat expansion in Friedreich ataxia interferes with transcription and may be associated with an unusual DNA structure. *Am J Hum Genet*. 1998 Jan;62(1):111–21
4. Babcock M, de Silva D, Oaks R, Davis-Kaplan S, Jiralerspong S, Montermini L, Pandolfo M, Kaplan J. Regulation of mitochondrial iron accumulation by Yfh1p, a putative homolog of frataxin. *Science*. 1997 Jun 13;276(5319):1709–12

5. Kremer EJ, Pritchard M, Lynch M, Yu S, Holman K, Baker E, Warren ST, Schlessinger D, Sutherland GR, Richards RI. Mapping of DNA instability at the fragile X to a trinucleotide repeat sequence p(CCG)n. *Science*. 1991 Jun 21;252(5013):1711–4
6. Verkerk AJ, Pieretti M, Sutcliffe JS, Fu YH, Kuhl DP, Pizzuti A, Reiner O, Richards S, Victoria MF, Zhang FP, et al. Identification of a gene (FMR-1) containing a CGG repeat coincident with a breakpoint cluster region exhibiting length variation in fragile X syndrome. *Cell*. 1991 May 31;65(5):905–14
7. Pieretti M, Zhang FP, Fu YH, Warren ST, Oostra BA, Caskey CT, Nelson DL. Absence of expression of the FMR-1 gene in fragile X syndrome. *Cell*. 1991 Aug 23;66(4):817–22
8. Qin M, Kang J, Burlin TV, Jiang C, Smith CB. Postadolescent changes in regional cerebral protein synthesis: an in vivo study in the FMR1 null mouse. *J Neurosci*. 2005 May 18;25(20):5087–95
9. Reviewed in Echeverria GV, Cooper TA. RNA-binding proteins in microsatellite expansion disorders: mediators of RNA toxicity. *Brain Res*. 2012 Jun 26;1462:100–11

Chapter 3

1. http://www.genome.gov/11006943
2. Unless otherwise stated, the majority of the information in this chapter is from the edition of *Nature* published on 15th February 2001 which contained the data and analyses from the publicly funded consortium. The major reference *is Initial sequencing and analysis of the human genome*, authored by the International Human Genome Sequencing Consortium. Readers may also find the accompanying commentaries in the same issue of interest.
3. http://partners.nytimes.com/library/national/science/062700sci-genome-text.html
4. http://news.bbc.co.uk/1/hi/sci/tech/807126.stm
5. http://news.bbc.co.uk/1/hi/sci/tech/807126.stm
6. http://www.genome.gov/sequencingcosts/
7. http://www.wired.co.uk/news/archive/2014-01/15/1000-dollar-genome
8. For a fascinating case history, see Gura, *Nature*, 2012, Volume 483, pp20–22

9. http://www.cancerresearchuk.org/cancer-help/about-cancer/treatment/cancer-drugs/Crizotinib/crizotinib
10. https://genographic.nationalgeographic.com/human-journey/
11. http://publications.nigms.nih.gov/insidelifescience/genetics-numbers.html
12. Aparicio et al. Whole-genome shotgun assembly and analysis of the genome of Fugu rubripes. *Science*. 2002 Aug 23;297(5585):1301–10
13. Baltimore D. Our genome unveiled. *Nature*. 2001 Feb 15;409(6822):814–6.
14. Data from the American Cancer Society http://www.cancer.org/cancer/skincancer-melanoma/detailedguide/melanoma-skin-cancer-key-statistics

Chapter 4

1. Unless otherwise stated, the majority of the information in this chapter is from the edition of *Nature* published on 15th February 2001 which contained the data and analyses from the publicly funded consortium. The major reference is *Initial sequencing and analysis of the human genome*, authored by the International Human Genome Sequencing Consortium. The accompanying commentaries by David Baltimore and by Li et al in the same issue are also of interest, and rather more accessible in style and content.
2. Vlangos CN, Siuniak AN, Robinson D, Chinnaiyan AM, Lyons RH Jr, Cavalcoli JD, Keegan CE. Next-generation sequencing identifies the Danforth's short tail mouse mutation as a retrotransposon insertion affecting Ptf1a expression. *PLoS Genet*. 2013;9(2):e1003205
3. Bogdanik LP, Chapman HD, Miers KE, Serreze DV, Burgess RW. A MusD retrotransposon insertion in the mouse Slc6a5 gene causes alterations in neuromuscular junction maturation and behavioral phenotypes. *PLoS One*. 2012;7(1):e30217
4. Schneuwly S, Klemenz R, Gehring WJ. Redesigning the body plan of Drosophila by ectopic expression of the homoeotic gene Antennapedia. *Nature*. 1987 Feb 26–Mar 4;325(6107):816–8
5. Mortlock DP, Post LC, Innis JW. The molecular basis of hypodactyly (Hd): a deletion in Hoxa 13 leads to arrest of digital arch formation. *Nat Genet*. 1996 Jul;13(3):284–9

6. Rowe HM, Jakobsson J, Mesnard D, Rougemont J, Reynard S, Aktas T, Maillard PV, Layard-Liesching H, Verp S, Marquis J, Spitz F, Constam DB, Trono D. KAP1 controls endogenous retroviruses in embryonic stem cells. *Nature*. 2010 Jan 14;463 (7278):237–40
7. Young GR, Eksmond U, Salcedo R, Alexopoulou L, Stoye JP, Kassiotis G. Resurrection of endogenous retroviruses in antibody-deficient mice. *Nature*. 2012 Nov 29;491(7426):774–8
8. http://www.emedicinehealth.com/heart_and_lung_transplant/article_em.htm
9. For an interesting recent review of the field of xenotransplanation, see Cooper DK. A brief history of cross-species organ transplantation. *Proc (Bayl Univ Med Cent)*. 2012 Jan;25(1):49–57
10. Patience C, Takeuchi Y, Weiss RA. Infection of human cells by an endogenous retrovirus of pigs. *Nat Med*. 1997 Mar;3(3):282–6
11. Di Nicuolo G, D'Alessandro A, Andria B, Scuderi V, Scognamiglio M, Tammaro A, Mancini A, Cozzolino S, Di Florio E, Bracco A, Calise F, Chamuleau RA. Long-term absence of porcine endogenous retrovirus infection in chronically immunosuppressed patients after treatment with the porcine cell-based Academic Medical Center bioartificial liver. *Xenotransplantation*. 2010 Nov–Dec;17(6):431–9
12. For a useful recent review of the effects of segmental duplication, including abnormal crossing-over, see Rudd MK, Keene J, Bunke B, Kaminsky EB, Adam MP, Mulle JG, Ledbetter DH, Martin CL. Segmental duplications mediate novel, clinically relevant chromosome rearrangements. *Hum Mol Genet*. 2009 Aug 15;18(16):2957–62
13. For more information on this condition and its causes, see http://www.ninds.nih.gov/disorders/charcot_marie_tooth/detail_charcot_marie_tooth.htm
14. For more information on this condition and its causes, see http://www.nlm.nih.gov/medlineplus/ency/article/001116.htm
15. Mombaerts P. The human repertoire of odorant receptor genes and pseudogenes. *Annu Rev Genomics Hum Genet*. 2001;2:493–510
16. http://www.innocenceproject.org/know/ retrieved 1 January 2014

Chapter 5

1. Gross takings as cited by http://www.imdb.com

2. Reviewed in Boxer LM, Dang CV. Translocations involving c-myc and c-myc function. *Oncogene*. 2001 Sep 20(40):5595–610
3. Moyzis RK, Buckingham JM, Cram LS, Dani M, Deaven LL, Jones MD, Meyne J, Ratliff RL, Wu JR. A highly conserved repetitive DNA sequence, (TTAGGG)n, present at the telomeres of human chromosomes. *Proc Natl Acad Sci U S A*. 1988 Sep;85(18):6622–6
4. Vaziri H, Schächter F, Uchida I, Wei L, Zhu X, Effros R, Cohen D, Harley CB. Loss of telomeric DNA during aging of normal and trisomy 21 human lymphocytes. *Am J Hum Genet*. 1993 Apr;52(4):661–7
5. Hayflick L, Moorhead PS. The serial cultivation of human diploid cell strains. *Exp Cell Res*. 1961 Dec;25:585–621
6. Harley CB, Futcher AB, Greider CW. Telomeres shorten during ageing of human fibroblasts. *Nature*. 1990 May 31;345(6274):458–60
7. Bodnar AG, Ouellette M, Frolkis M, Holt SE, Chiu CP, Morin GB, Harley CB, Shay JW, Lichtsteiner S, Wright WE. Extension of life-span by introduction of telomerase into normal human cells. *Science*. 1998 Jan 16;279(5349):349–52
8. There is a useful discussion of this problem in Armanios M, Blackburn EH. The telomere syndromes. *Nat Rev Genet*. 2012 Oct;13(10):693–704
9. Armanios M, Blackburn EH. The telomere syndromes. *Nat Rev Genet*. 2012 Oct;13(10):693–704 provides a useful overview.
10. Wright WE, Piatyszek MA, Rainey WE, Byrd W, Shay JW. Telomerase activity in human germline and embryonic tissues and cells. *Dev Genet*. 1996;18(2):173–9
11. Kim NW, Piatyszek MA, Prowse KR, Harley CB, West MD, Ho PL, Coviello GM, Wright WE, Weinrich SL, Shay JW. Specific association of human telomerase activity with immortal cells and cancer. *Science*. 1994 Dec 23;266(5193):2011–5
12. http://www.nlm.nih.gov/medlineplus/ency/anatomyvideos/000104.htm
13. Chiu CP, Dragowska W, Kim NW, Vaziri H, Yui J, Thomas TE, Harley CB, Lansdorp PM. Differential expression of telomerase activity in hematopoietic progenitors from adult human bone marrow. *Stem Cells*. 1996 Mar;14(2):239–48
14. Vaziri H, Dragowska W, Allsopp RC, Thomas TE, Harley CB, Lansdorp PM. Evidence for a mitotic clock in human hematopoietic

stem cells: loss of telomeric DNA with age. *Proc Natl Acad Sci U S A.* 1994 Oct 11;91(21):9857–60

15. Armanios M, Blackburn EH. The telomere syndromes. *Nat Rev Genet.* 2012 Oct;13(10):693–704
16. Armanios M, Blackburn EH. The telomere syndromes. *Nat Rev Genet.* 2012 Oct;13(10):693–704
17. For an excellent clinical description, and useful pictures, see Calado RT, Young NS. Telomere diseases. *N Engl J Med.* 2009 Dec 10;361(24):2353–65
18. Alder JK, Chen JJ, Lancaster L, Danoff S, Su SC, Cogan JD, Vulto I, Xie M, Qi X, Tuder RM, Phillips JA 3rd, Lansdorp PM, Loyd JE, Armanios MY. Short telomeres are a risk factor for idiopathic pulmonary fibrosis. *Proc Natl Acad Sci U S A.* 2008 Sep 2;105(35):13051–6
19. Armanios MY, Chen JJ, Cogan JD, Alder JK, Ingersoll RG, Markin C, Lawson WE, Xie M, Vulto I, Phillips JA 3rd, Lansdorp PM, Greider CW, Loyd JE. Telomerase mutations in families with idiopathic pulmonary fibrosis. *N Engl J Med.* 2007 Mar 29;356(13):1317–26
20. Tsakiri KD, Cronkhite JT, Kuan PJ, Xing C, Raghu G, Weissler JC, Rosenblatt RL, Shay JW, Garcia CK. Adult-onset pulmonary fibrosis caused by mutations in telomerase. *Proc Natl Acad Sci U S A.* 2007 May 1;104(18):7552–7
21. Cronkhite JT, Xing C, Raghu G, Chin KM, Torres F, Rosenblatt RL, Garcia CK. Telomere shortening in familial and sporadic pulmonary fibrosis. *Am J Respir Crit Care Med.* 2008 Oct 1;178(7):729–37
22. For a useful description see http://www.patient.co.uk/doctor/aplastic-anaemia
23. de la Fuente J, Dokal I. Dyskeratosis congenita: advances in the understanding of the telomerase defect and the role of stem cell transplantation. *Pediatr Transplant.* 2007 Sep;11(6):584–94
24. Armanios M, Chen JL, Chang YP, Brodsky RA, Hawkins A, Griffin CA, Eshleman JR, Cohen AR, Chakravarti A, Hamosh A, Greider CW. Haploinsufficiency of telomerase reverse transcriptase leads to anticipation in autosomal dominant dyskeratosis congenita. *Proc Natl Acad Sci U S A.* 2005 Nov 1;102(44):15960–4
25. http://www.who.int/mediacentre/factsheets/fs339/en/
26. Alder JK, Guo N, Kembou F, Parry EM, Anderson CJ, Gorgy AI,

Walsh MF, Sussan T, Biswal S, Mitzner W, Tuder RM, Armanios M. Telomere length is a determinant of emphysema susceptibility. *Am J Respir Crit Care Med*. 2011 Oct 15;184(8):904–12

27. Cited in Sahin E, Depinho RA. Linking functional decline of telomeres, mitochondria and stem cells during ageing. *Nature*. 2010 Mar 25;464(7288):520–8
28. Statistical factsheet from the American Heart Association on Older Americans & Cardiovascular Diseases, 2013 update
29. http://www.rcpsych.ac.uk/healthadvice/problemsdisorders/depressioninolderadults.aspx
30. Valdes AM, Andrew T, Gardner JP, Kimura M, Oelsner E, Cherkas LF, Aviv A, Spector TD. Obesity, cigarette smoking, and telomere length in women. *Lancet*. 2005 Aug 20–26;366(9486):662–4
31. Cawthon RM, Smith KR, O'Brien E, Sivatchenko A, Kerber RA. Association between telomere length in blood and mortality in people aged 60 years or older. *Lancet*. 2003 Feb 1;361(9355):393–5
32. Fitzpatrick AL, Kronmal RA, Kimura M, Gardner JP, Psaty BM, Jenny NS, Tracy RP, Hardikar S, Aviv A. Leukocyte telomere length and mortality in the Cardiovascular Health Study. *J Gerontol A Biol Sci Med Sci*. 2011 Apr;66(4):421–9
33. Atzmon G, Cho M, Cawthon RM, Budagov T, Katz M, Yang X, Siegel G, Bergman A, Huffman DM, Schechter CB, Wright WE, Shay JW, Barzilai N, Govindaraju DR, Suh Y. Evolution in health and medicine Sackler colloquium: Genetic variation in human telomerase is associated with telomere length in Ashkenazi centenarians. *Proc Natl Acad Sci U S A*. 2010 Jan 26;107 Suppl 1:1710–7
34. Segerstrom SC, Miller GE. Psychological stress and the human immune system: a meta-analytic study of 30 years of inquiry. *Psychol Bull*. 2004 Jul;130(4):601–30
35. Epel ES, Blackburn EH, Lin J, Dhabhar FS, Adler NE, Morrow JD, Cawthon RM. Accelerated telomere shortening in response to life stress. *Proc Natl Acad Sci U S A*. 2004 Dec 7;101(49):17312–5
36. http://www.who.int/mediacentre/factsheets/fs311/en/index.html
37. For a useful introduction to this field, see Tennen RI, Chua KF. Chromatin regulation and genome maintenance by mammalian SIRT6. *Trends Biochem Sci*. 2011 Jan;36(1):39–46

38. Valdes AM, Andrew T, Gardner JP, Kimura M, Oelsner E, Cherkas LF, Aviv A, Spector TD. Obesity, cigarette smoking, and telomere length in women. *Lancet*. 2005 Aug 20–26;366(9486):662–4
39. UNFPA report on Ageing in The Twenty-First Century, 2012
40. Jennings BJ, Ozanne SE, Dorling MW, Hales CN. Early growth determines longevity in male rats and may be related to telomere shortening in the kidney. *FEBS Lett*. 1999 Apr 1;448(1):4–8

Chapter 6

1. From *The King and I*, 1956, screenplay by Ernest Lehman, 20th Century Fox
2. A good overview of the types of centromeres in the different arms of the evolutionary tree can be found in Ogiyama Y, Ishii K. The smooth and stable operation of centromeres. *Genes Genet Syst*. 2012;87(2):63–73
3. For a useful review, see Verdaasdonk JS, Bloom K. Centromeres: unique chromatin structures that drive chromosome segregation. *Nat Rev Mol Cell Biol*. 2011 May;12(5):320–32
4. Palmer DK, O'Day K, Wener MH, Andrews BS, Margolis RL. A 17-kD centromere protein (CENP-A) copurifies with nucleosome core particles and with histones. *J Cell Biol*. 1987 Apr;104(4):805–15
5. Takahashi K, Chen ES, Yanagida M. Requirement of Mis6 centromere connector for localizing a CENP-A-like protein in fission yeast. *Science*. 2000 Jun 23;288(5474):2215–9
6. Blower MD, Karpen GH. The role of Drosophila CID in kinetochore formation, cell-cycle progression and heterochromatin interactions. *Nat Cell Biol*. 2001 Aug;3(8):730–9
7. Hori T, Amano M, Suzuki A, Backer CB, Welburn JP, Dong Y, McEwen BF, Shang WH, Suzuki E, Okawa K, Cheeseman IM, Fukagawa T. CCAN makes multiple contacts with centromeric DNA to provide distinct pathways to the outer kinetochore. *Cell*. 2008 Dec 12;135(6):1039–52
8. Heun P, Erhardt S, Blower MD, Weiss S, Skora AD, Karpen GH. Mislocalization of the Drosophila centromere-specific histone CID promotes formation of functional ectopic kinetochores. *Dev Cell*. 2006 Mar;10(3):303–15.
9. Van Hooser AA, Ouspenski II, Gregson HC, Starr DA, Yen TJ,

Goldberg ML, Yokomori K, Earnshaw WC, Sullivan KF, Brinkley BR. Specification of kinetochore-forming chromatin by the histone H3 variant CENP-A. *J Cell Sci.* 2001 Oct;114(Pt 19):3529–42

10. Zuccolo M, Alves A, Galy V, Bolhy S, Formstecher E, Racine V, Sibarita JB, Fukagawa T, Shiekhattar R, Yen T, Doye V. The human Nup107-160 nuclear pore subcomplex contributes to proper kinetochore functions. *EMBO J.* 2007 Apr 4;26(7):1853–64
11. Palmer DK, O'Day K, Wener MH, Andrews BS, Margolis RL. A 17-kD centromere protein (CENP-A) copurifies with nucleosome core particles and with histones. *J Cell Biol.* 1987 Apr;104(4):805–15
12. Sekulic N, Bassett EA, Rogers DJ, Black BE. The structure of (CENP-A-H4)(2) reveals physical features that mark centromeres. *Nature.* 2010 Sep 16;467(7313):347–51
13. Warburton PE, Cooke CA, Bourassa S, Vafa O, Sullivan BA, Stetten G, Gimelli G, Warburton D, Tyler-Smith C, Sullivan KF, Poirier GG, Earnshaw WC. Immunolocalization of CENP-A suggests a distinct nucleosome structure at the inner kinetochore plate of active centromeres. *Curr Biol.* 1997 Nov 1;7(11):901–4
14. For a very good analysis of this model, see Sekulic N, Black BE. Molecular underpinnings of centromere identity and maintenance. *Trends Biochem Sci.* 2012 Jun;37(6):220–9
15. If you are interested in learning more about the details of this process, and the epigenetic modifications involved, see González-Barrios R, Soto-Reyes E, Herrera LA. Assembling pieces of the centromere epigenetics puzzle. *Epigenetics.* 2012 Jan 1;7(1):3–13
16. From the song 'Something Good' in the movie version of *The Sound of Music*, 1965, 20th Century Fox
17. A particularly important protein in this respect is call HJURP, and more information can be found in Sekulic N, Black BE. Molecular underpinnings of centromere identity and maintenance. *Trends Biochem Sci.* 2012 Jun;37(6):220–9
18. Palmer DK, O'Day K, Margolis RL. The centromere specific histone CENP-A is selectively retained in discrete foci in mammalian sperm nuclei. *Chromosoma.* 1990 Dec;100(1):32–6
19. Schiff PB, Fant J, Horwitz SB. Promotion of microtubule assembly in vitro by taxol. *Nature.* 1979 Feb 22;277(5698):665–7

20. http://www.cancerresearchuk.org/cancer-help/about-cancer/treatment/cancer-drugs/paclitaxel
21. Figure quoted in Rajagopalan H, Lengauer C. Aneuploidy and cancer. *Nature*. 2004 Nov 18;432(7015):338–41
22. For a review of this issue, see Pfau SJ, Amon A. Chromosomal instability and aneuploidy in cancer: from yeast to man. *EMBO Rep*. 2012 Jun 1;13(6):515–27
23. Rehen SK, Yung YC, McCreight MP, Kaushal D, Yang AH, Almeida BS, Kingsbury MA, Cabral KM, McConnell MJ, Anliker B, Fontanoz M, Chun J. Constitutional aneuploidy in the normal human brain. *J Neurosci*. 2005 Mar 2;25(9):2176–80
24. Rehen SK, McConnell MJ, Kaushal D, Kingsbury MA, Yang AH, Chun J. Chromosomal variation in neurons of the developing and adult mammalian nervous system. *Proc Natl Acad Sci U S A*. 2001 Nov 6;98(23):13361–6
25. Kingsbury MA, Friedman B, McConnell MJ, Rehen SK, Yang AH, Kaushal D, Chun J. Aneuploid neurons are functionally active and integrated into brain circuitry. *Proc Natl Acad Sci U S A*. 2005 Apr 26;102(17):6143–7
26. Melchiorri C, Chieco P, Zedda AI, Coni P, Ledda-Columbano GM, Columbano A. Ploidy and nuclearity of rat hepatocytes after compensatory regeneration or mitogen-induced liver growth. *Carcinogenesis*. 1993 Sep;14(9):1825–30
27. For an extraordinary account of the ill-tempered controversy over who exactly identified the cause of Down's Syndrome, which is still raging after 50 years, see http://www.nature.com/news/down-s-syndrome-discovery-dispute-resurfaces-in-france-1.14690
28. For more information on the medical and social aspects of Down's Syndrome there are a large number of patient advocacy groups such as http://www.downs-syndrome.org.uk/
29. http://www.nhs.uk/conditions/edwards-syndrome/Pages/Introduction.aspx
30. http://www.cafamily.org.uk/medical-information/conditions/p/patau-syndrome/
31. Toner JP, Grainger DA, Frazier LM. Clinical outcomes among recipients of donated eggs: an analysis of the U.S. national experience, 1996–1998. *Fertil Steril*. 2002 Nov;78(5):1038–45

Chapter 7

1. Statistical Bulletin from the Office for National Statistics, 8 August 2013 Annual Mid-year Population Estimates, 2011 and 2012
2. The publication that demonstrated the importance of this gene is Berta P, Hawkins JR, Sinclair AH, Taylor A, Griffiths BL, Goodfellow PN, Fellous M. Genetic evidence equating SRY and the testis-determining factor. *Nature*. 1990 Nov 29;348(6300):448–50
3. Yamauchi Y, Riel JM, Stoytcheva Z, Ward MA. Two Y genes can replace the entire Y chromosome for assisted reproduction in the mouse. *Science*. 2014 Jan 3;343(6166):69–72
4. Ross MT et al., The DNA sequence of the human X chromosome. *Nature*. 2005 Mar 17;434(7031):325–37
5. Brown CJ, Lafreniere RG, Powers VE, Sebastio G, Ballabio A, Pettigrew AL, Ledbetter DH, Levy E, Craig IW, Willard HF. Localization of the X inactivation centre on the human X chromosome in Xq13. *Nature*. 1991 Jan 3;349(6304):82–4
6. Brown CJ, Ballabio A, Rupert JL, Lafreniere RG, Grompe M, Tonlorenzi R, Willard HF. A gene from the region of the human X inactivation centre is expressed exclusively from the inactive X chromosome. *Nature*. 1991 Jan 3;349(6304):38–44
7. Brown CJ, Hendrich BD, Rupert JL, Lafrenière RG, Xing Y, Lawrence J, Willard HF. The human XIST gene: analysis of a 17 kb inactive X-specific RNA that contains conserved repeats and is highly localized within the nucleus. *Cell*. 1992 Oct 30;71(3):527–42
8. Brockdorff N, Ashworth A, Kay GF, McCabe VM, Norris DP, Cooper PJ, Swift S, Rastan S. The product of the mouse Xist gene is a 15 kb inactive X-specific transcript containing no conserved ORF and located in the nucleus. *Cell*. 1992 Oct 30;71(3):515–26
9. Lee JT, Strauss WM, Dausman JA, Jaenisch R. A 450 kb transgene displays properties of the mammalian X-inactivation center. *Cell*. 1996 Jul 12;86(1):83–94
10. For a comprehensive review of this process, see Lee JT. The X as model for RNA's niche in epigenomic regulation. *Cold Spring Harb Perspect Biol*. 2010 Sep;2(9):a003749
11. Xu N, Tsai CL, Lee JT. Transient homologous chromosome

pairing marks the onset of X inactivation. *Science*. 2006 Feb 24;311(5764):1149–52
12. For a fascinating précis of the spread of haemophilia through the European royal families, see http://www.hemophilia.org/NHFWeb/MainPgs/MainNHF.aspx?menuid=178&contentid=6
13. For more information on this condition see http://www.nhs.uk/conditions/Rett-syndrome/Pages/Introduction.aspx
14. Amir RE, Van den Veyver IB, Wan M, Tran CQ, Francke U, Zoghbi HY. Rett syndrome is caused by mutations in X-linked MECP2, encoding methyl-CpG-binding protein 2. *Nat Genet*. 1999 Oct;23(2):185–8
15. For more information on this condition, see http://www.nlm.nih.gov/medlineplus/ency/article/000705.htm
16. Hoffman EP, Brown RH Jr, Kunkel LM. Dystrophin: the protein product of the Duchenne muscular dystrophy locus. *Cell*. 1987 Dec 24;51(6):919–28
17. Pena SD, Karpati G, Carpenter S, Fraser FC. The clinical consequences of X-chromosome inactivation: Duchenne muscular dystrophy in one of monozygotic twins. *J Neurol Sci*. 1987 Jul;79(3):337–44
18. Shin T, Kraemer D, Pryor J, Liu L, Rugila J, Howe L, Buck S, Murphy K, Lyons L, Westhusin M. A cat cloned by nuclear transplantation. *Nature*. 2002 Feb 21;415(6874):859

Chapter 8

1. Schmitt AM, Chang HY. Gene regulation: Long RNAs wire up cancer growth. *Nature*. 2013 Aug 29;500(7464):536–7
2. Volders PJ, Helsens K, Wang X, Menten B, Martens L, Gevaert K, Vandesompele J, Mestdagh P. LNCipedia: a database for annotated human long-noncoding RNA transcript sequences and structures. *Nucleic Acids Res*. 2013 Jan;41(Database issue):D246–51
3. ENCODE Project Consortium, Bernstein BE, Birney E, Dunham I, Green ED, Gunter C, Snyder M. An integrated encyclopedia of DNA elements in the human genome. *Nature*. 2012 Sep 6;489(7414):57–74
4. Tay Y, Rinn J, Pandolfi PP. The multilayered complexity of ceRNA crosstalk and competition. *Nature*. 2014 Jan 16;505(7483):344–52

5. Derrien T, Johnson R, Bussotti G, Tanzer A, Djebali S, Tilgner H, Guernec G, Martin D, Merkel A, Knowles DG, Lagarde J, Veeravalli L, Ruan X, Ruan Y, Lassmann T, Carninci P, Brown JB, Lipovich L, Gonzalez JM, Thomas M, Davis CA, Shiekhattar R, Gingeras TR, Hubbard TJ, Notredame C, Harrow J, Guigó R. The GENCODE v7 catalog of human long noncoding RNAs: analysis of their gene structure, evolution, and expression. *Genome Res*. 2012 Sep;22(9):1775–89
6. Ulitsky I, Shkumatava A, Jan CH, Sive H, Bartel DP. Conserved function of lincRNAs in vertebrate embryonic development despite rapid sequence evolution. *Cell*. 2011 Dec 23;147(7):1537–50
7. Cabili MN, Trapnell C, Goff L, Koziol M, Tazon-Vega B, Regev A, Rinn JL. Integrative annotation of human large intergenic noncoding RNAs reveals global properties and specific subclasses. *Genes Dev*. 2011 Sep 15;25(18):1915–27
8. Church DM, Goodstadt L, Hillier LW, Zody MC, Goldstein S, She X, Bult CJ, Agarwala R, Cherry JL, DiCuccio M, Hlavina W, Kapustin Y, Meric P, Maglott D, Birtle Z, Marques AC, Graves T, Zhou S, Teague B, Potamousis K, Churas C, Place M, Herschleb J, Runnheim R, Forrest D, Amos-Landgraf J, Schwartz DC, Cheng Z, Lindblad-Toh K, Eichler EE, Ponting CP; Mouse Genome Sequencing Consortium. Lineage-specific biology revealed by a finished genome assembly of the mouse. *PLoS Biol*. 2009 May 5;7(5):e1000112
9. Necsulea A, Soumillon M, Warnefors M, Liechti A, Daish T, Zeller U, Baker JC, Grützner F, Kaessmann H. The evolution of long-noncoding RNA repertoires and expression patterns in tetrapods. *Nature*. 2014 Jan 30;505(7485):635–40
10. Wahlestedt C. Targeting long non-coding RNA to therapeutically upregulate gene expression. *Nat Rev Drug Discov*. 2013 Jun;12(6):433–46
11. Mercer TR, Dinger ME, Sunkin SM, Mehler MF, Mattick JS. Specific expression of long noncoding RNAs in the mouse brain. *Proc Natl Acad Sci U S A*. 2008 Jan 15;105(2):716–21
12. For a very useful review of this class and how it fits into the wider long non-coding RNA landscape, see Ulitsky I, Bartel DP. lincRNAs: genomics, evolution, and mechanisms. *Cell*. 2013 Jul 3;154(1):26–46

13. Guttman M, Donaghey J, Carey BW, Garber M, Grenier JK, Munson G, Young G, Lucas AB, Ach R, Bruhn L, Yang X, Amit I, Meissner A, Regev A, Rinn JL, Root DE, Lander ES. lincRNAs act in the circuitry controlling pluripotency and differentiation. *Nature*. 2011 Aug 28;477(7364):295–300
14. Wang KC, Yang YW, Liu B, Sanyal A, Corces-Zimmerman R, Chen Y, Lajoie BR, Protacio A, Flynn RA, Gupta RA, Wysocka J, Lei M, Dekker J, Helms JA, Chang HY. A long noncoding RNA maintains active chromatin to coordinate homeotic gene expression. *Nature*. 2011 Apr 7;472(7341):120–4
15. Li L, Liu B, Wapinski OL, Tsai MC, Qu K, Zhang J, Carlson JC, Lin M, Fang F, Gupta RA, Helms JA, Chang HY. Targeted disruption of Hotair leads to homeotic transformation and gene derepression. *Cell Rep*. 2013 Oct 17;5(1):3–12
16. Du Z, Fei T, Verhaak RG, Su Z, Zhang Y, Brown M, Chen Y, Liu XS. Integrative genomic analyses reveal clinically relevant long noncoding RNAs in human cancer. *Nat Struct Mol Biol*. 2013 Jul;20(7):908–13
17. For a useful review of this area, see Cheetham SW, Gruhl F, Mattick JS, Dinger ME. Long noncoding RNAs and the genetics of cancer. *Br J Cancer*. 2013 Jun 25;108(12):2419–25
18. Yap KL, Li S, Muñoz-Cabello AM, Raguz S, Zeng L, Mujtaba S, Gil J, Walsh MJ, Zhou MM. Molecular interplay of the noncoding RNA ANRIL and methylated histone H3 lysine 27 by polycomb CBX7 in transcriptional silencing of INK4a. *Mol Cell*. 2010 Jun 11;38(5):662–74
19. Kotake Y, Nakagawa T, Kitagawa K, Suzuki S, Liu N, Kitagawa M, Xiong Y. Long non-coding RNA ANRIL is required for the PRC2 recruitment to and silencing of p15(INK4B) tumor suppressor gene. *Oncogene*. 2011 Apr 21;30(16):1956–62
20. Yang Z, Zhou L, Wu LM, Lai MC, Xie HY, Zhang F, Zheng SS. Overexpression of long non-coding RNA HOTAIR predicts tumor recurrence in hepatocellular carcinoma patients following liver transplantation. *Ann Surg Oncol*. 2011 May;18(5):1243–50
21. Ishibashi M, Kogo R, Shibata K, Sawada G, Takahashi Y, Kurashige J, Akiyoshi S, Sasaki S, Iwaya T, Sudo T, Sugimachi K, Mimori K, Wakabayashi G, Mori M. Clinical significance of the expression of long

non-coding RNA HOTAIR in primary hepatocellular carcinoma. *Oncol Rep*. 2013 Mar;29(3):946–50

22. Kim K, Jutooru I, Chadalapaka G, Johnson G, Frank J, Burghardt R, Kim S, Safe S. HOTAIR is a negative prognostic factor and exhibits pro-oncogenic activity in pancreatic cancer. *Oncogene*. 2013 Mar 8;32(13):1616–25
23. Gupta RA, Shah N, Wang KC, Kim J, Horlings HM, Wong DJ, Tsai MC, Hung T, Argani P, Rinn JL, Wang Y, Brzoska P, Kong B, Li R, West RB, van de Vijver MJ, Sukumar S, Chang HY. Long non-coding RNA HOTAIR reprograms chromatin state to promote cancer metastasis. *Nature*. 2010 Apr 15;464(7291):1071–6
24. Yang L, Lin C, Jin C, Yang JC, Tanasa B, Li W, Merkurjev D, Ohgi KA, Meng D, Zhang J, Evans CP, Rosenfeld MG. Long-noncoding RNA-dependent mechanisms of androgen-receptor-regulated gene activation programs. *Nature*. 2013 Aug 29;500(7464):598–602
25. Prensner JR, Iyer MK, Sahu A, Asangani IA, Cao Q, Patel L, Vergara IA, Davicioni E, Erho N, Ghadessi M, Jenkins RB, Triche TJ, Malik R, Bedenis R, McGregor N, Ma T, Chen W, Han S, Jing X, Cao X, Wang X, Chandler B, Yan W, Siddiqui J, Kunju LP, Dhanasekaran SM, Pienta KJ, Feng FY, Chinnaiyan AM. The long noncoding RNA SChLAP1 promotes aggressive prostate cancer and antagonizes the SWI/SNF complex. *Nat Genet*. 2013 Nov;45(11):1392–8
26. Necsulea A, Soumillon M, Warnefors M, Liechti A, Daish T, Zeller U, Baker JC, Grützner F, Kaessmann H. The evolution of long-noncoding RNA repertoires and expression patterns in tetrapods. *Nature*. 2014 Jan 30;505(7485):635–40
27. For an interesting critique of this issue, see Fatica A, Bozzoni I. Long non-coding RNAs: new players in cell differentiation and development. *Nat Rev Genet*. 2014 Jan;15(1):7–21
28. Bernard D, Prasanth KV, Tripathi V, Colasse S, Nakamura T, Xuan Z, Zhang MQ, Sedel F, Jourdren L, Coulpier F, Triller A, Spector DL, Bessis A. A long nuclear-retained non-coding RNA regulates synaptogenesis by modulating gene expression. *EMBO J*. 2010 Sep 15;29(18):3082–93
29. Pollard KS, Salama SR, Lambert N, Lambot MA, Coppens S, Pedersen JS, Katzman S, King B, Onodera C, Siepel A, Kern AD, Dehay C, Igel

H, Ares M Jr, Vanderhaeghen P, Haussler D. An RNA gene expressed during cortical development evolved rapidly in humans. *Nature*. 2006 Sep 14;443(7108):167–72

30. http://www.who.int/mental_health/publications/dementia_report_2012/en/
31. Faghihi MA, Modarresi F, Khalil AM, Wood DE, Sahagan BG, Morgan TE, Finch CE, St Laurent G 3rd, Kenny PJ, Wahlestedt C. Expression of a noncoding RNA is elevated in Alzheimer's disease and drives rapid feed-forward regulation of beta-secretase. *Nat Med*. 2008 Jul;14(7):723–30
32. Modarresi F, Faghihi MA, Patel NS, Sahagan BG, Wahlestedt C, Lopez-Toledano MA. Knockdown of BACE1-AS Nonprotein-Coding Transcript Modulates Beta-Amyloid-Related Hippocampal Neurogenesis. *Int J Alzheimers Dis*. 2011;2011:929042
33. Zhao X, Tang Z, Zhang H, Atianjoh FE, Zhao JY, Liang L, Wang W, Guan X, Kao SC, Tiwari V, Gao YJ, Hoffman PN, Cui H, Li M, Dong X, Tao YX. A long noncoding RNA contributes to neuropathic pain by silencing Kcna2 in primary afferent neurons. *Nat Neurosci*. 2013 Aug;16(8):1024–31
34. For a useful review, see for example Wahlestedt C. Targeting long non-coding RNA to therapeutically upregulate gene expression. *Nat Rev Drug Discov*. 2013 Jun;12(6):433–46
35. Bird A. Genome biology: not drowning but waving. *Cell*. 2013 Aug 29;154(5):951–2

Chapter 9

1. If you want to learn more about this topic, have a read of my first book, *The Epigenetics Revolution*.
2. Guttman M, Donaghey J, Carey BW, Garber M, Grenier JK, Munson G, Young G, Lucas AB, Ach R, Bruhn L, Yang X, Amit I, Meissner A, Regev A, Rinn JL, Root DE, Lander ES. lincRNAs act in the circuitry controlling pluripotency and differentiation. *Nature*. 2011 Aug 28;477(7364):295–300
3. Guil S, Soler M, Portela A, Carrère J, Fonalleras E, Gómez A, Villanueva A, Esteller M. Intronic RNAs mediate EZH2 regulation of epigenetic targets. *Nat Struct Mol Biol*. 2012 Jun 3;19(7):664–70

4. Varambally S, Dhanasekaran SM, Zhou M, Barrette TR, Kumar-Sinha C, Sanda MG, Ghosh D, Pienta KJ, Sewalt RG, Otte AP, Rubin MA, Chinnaiyan AM. The polycomb group protein EZH2 is involved in progression of prostate cancer. *Nature*. 2002 Oct 10;419(6907):624–9
5. Kleer CG, Cao Q, Varambally S, Shen R, Ota I, Tomlins SA, Ghosh D, Sewalt RG, Otte AP, Hayes DF, Sabel MS, Livant D, Weiss SJ, Rubin MA, Chinnaiyan AM. EZH2 is a marker of aggressive breast cancer and promotes neoplastic transformation of breast epithelial cells. *Proc Natl Acad Sci U S A*. 2003 Sep 30;100(20):11606–11.
6. Sneeringer CJ, Scott MP, Kuntz KW, Knutson SK, Pollock RM, Richon VM, Copeland RA. Coordinated activities of wild-type plus mutant EZH2 drive tumor-associated hypertrimethylation of lysine 27 on histone H3 (H3K27) in human B-cell lymphomas. *Proc Natl Acad Sci U S A*. 2010 Dec 7;107(49):20980–5
7. http://clinicaltrials.gov/ct2/show/NCT01897571?term=7438&rank=1
8. Kotake Y, Nakagawa T, Kitagawa K, Suzuki S, Liu N, Kitagawa M, Xiong Y. Long non-coding RNA ANRIL is required for the PRC2 recruitment to and silencing of p15(INK4B) tumor suppressor gene. *Oncogene*. 2011 Apr 21;30(16):1956–62
9. Tsai MC, Manor O, Wan Y, Mosammaparast N, Wang JK, Lan F, Shi Y, Segal E, Chang HY. Long noncoding RNA as modular scaffold of histone modification complexes. *Science*. 2010 Aug 6;329(5992):689–93
10. For a recent major paper on this see Davidovich C, Zheng L, Goodrich KJ, Cech TR. Promiscuous RNA binding by Polycomb repressive complex 2. *Nat Struct Mol Biol*. 2013 Nov;20(11):1250–7
11. For a slightly more accessible summary of the above paper, see Goff LA, Rinn JL. Poly-combing the genome for RNA. *Nat Struct Mol Biol*. 2013 Dec;20(12):1344–6
12. Di Ruscio A, Ebralidze AK, Benoukraf T, Amabile G, Goff LA, Terragni J, Figueroa ME, De Figueiredo Pontes LL, Alberich-Jorda M, Zhang P, Wu M, D'Alò F, Melnick A, Leone G, Ebralidze KK, Pradhan S, Rinn JL, Tenen DG. DNMT1-interacting RNAs block gene-specific DNA methylation. *Nature*. 2013 Nov 21;503(7476):371–6
13. For an overview of all the complex stages in this process see Froberg JE, Yang L, Lee JT. Guided by RNAs: X-inactivation as a model for long non-coding RNA function. *J Mol Biol*. 2013 Oct 9;425(19):3698–706

14. Froberg JE, Yang L, Lee JT. Guided by RNAs: X-inactivation as a model for long non-coding RNA function. *J Mol Biol*. 2013 Oct 9;425(19):3698–706
15. Michaud EJ, van Vugt MJ, Bultman SJ, Sweet HO, Davisson MT, Woychik RP. Differential expression of a new dominant agouti allele (Aiapy) is correlated with methylation state and is influenced by parental lineage. *Genes Dev*. 1994 Jun 15;8(12):1463–72

Chapter 10

1. For a contemporaneous review of the work see Surani MA, Barton SC, Norris ML. Experimental reconstruction of mouse eggs and embryos: an analysis of mammalian development. *Biol Reprod*. 1987 Feb;36(1):1–16
2. An online depository of imprinted mouse sequences can be found at http://www.mousebook.org/catalog.php?catalog=imprinting
3. For a useful review see Guenzl PM, Barlow DP. Macro long non-coding RNAs: a new layer of cis-regulatory information in the mammalian genome. *RNA Biol*. 2012 Jun;9(6):731–41
4. For a recent review of imprinting in marsupials see Graves JA, Renfree MB. Marsupials in the age of genomics. *Annu Rev Genomics Hum Genet*. 2013;14:393–420
5. Landers M, Bancescu DL, Le Meur E, Rougeulle C, Glatt-Deeley H, Brannan C, Muscatelli F, Lalande M. Regulation of the large (approximately 1000 kb) imprinted murine Ube3a antisense transcript by alternative exons upstream of Snurf/Snrpn. *Nucleic Acids Res*. 2004 Jun 29;32(11):3480–92
6. Terranova R, Yokobayashi S, Stadler MB, Otte AP, van Lohuizen M, Orkin SH, Peters AH. Polycomb group proteins Ezh2 and Rnf2 direct genomic contraction and imprinted repression in early mouse embryos. *Dev Cell*. 2008 Nov;15(5):668–79
7. Wagschal A, Sutherland HG, Woodfine K, Henckel A, Chebli K, Schulz R, Oakey RJ, Bickmore WA, Feil R. G9a histone methyltransferase contributes to imprinting in the mouse placenta. *Mol Cell Biol*. 2008 Feb;28(3):1104–13
8. Nagano T, Mitchell JA, Sanz LA, Pauler FM, Ferguson-Smith AC, Feil R, Fraser P. The Air noncoding RNA epigenetically silences

transcription by targeting G9a to chromatin. *Science*. 2008 Dec 12;322(5908):1717–20

9. Reviewed in Koerner MV, Pauler FM, Huang R, Barlow DP. The function of non-coding RNAs in genomic imprinting. *Development*. 2009 Jun;136(11):1771–83
10. Barlow DP. Methylation and imprinting: from host defense to gene regulation? *Science*. 1993 Apr 16;260(5106):309–10
11. Reviewed in Skaar DA, Li Y, Bernal AJ, Hoyo C, Murphy SK, Jirtle RL. The human imprintome: regulatory mechanisms, methods of ascertainment, and roles in disease susceptibility. *ILAR J*. 2012 Dec;53(3–4):341–58
12. A description of the actions of these proteins in the methylation of the maternal ICE can be found in Bourc'his D, Proudhon C. Sexual dimorphism in parental imprint ontogeny and contribution to embryonic development. *Mol Cell Endocrinol*. 2008 Jan 30;282(1–2):87–94
13. The paper that demonstrated the importance of this protein for maintaining the maternal imprint is Hirasawa R, Chiba H, Kaneda M, Tajima S, Li E, Jaenisch R, Sasaki H. Maternal and zygotic Dnmt1 are necessary and sufficient for the maintenance of DNA methylation imprints during preimplantation development. *Genes Dev*. 2008 Jun 15;22(12):1607–16
14. Reinhart B, Paoloni-Giacobino A, Chaillet JR. Specific differentially methylated domain sequences direct the maintenance of methylation at imprinted genes. *Mol Cell Biol*. 2006 Nov;26(22):8347–56
15. Skaar DA, Li Y, Bernal AJ, Hoyo C, Murphy SK, Jirtle RL. The human imprintome: regulatory mechanisms, methods of ascertainment, and roles in disease susceptibility. *ILAR J*. 2012 Dec;53(3–4):341–58
16. Kawahara M, Wu Q, Takahashi N, Morita S, Yamada K, Ito M, Ferguson-Smith AC, Kono T. High-frequency generation of viable mice from engineered bi-maternal embryos. *Nat Biotechnol*. 2007 Sep;25(9):1045–50
17. Reviewed in Fatica A, Bozzoni I. Long non-coding RNAs: new players in cell differentiation and development. *Nat Rev Genet*. 2014 Jan;15(1):7–21
18. For a review of this aspect, see Frost JM, Moore GE. The importance

of imprinting in the human placenta. *PLoS Genet*. 2010 Jul 1;6(7):e1001015
19. For a full description see http://omim.org/entry/176270
20. For a full description see http://omim.org/entry/105830
21. de Smith AJ, Purmann C, Walters RG, Ellis RJ, Holder SE, Van Haelst MM, Brady AF, Fairbrother UL, Dattani M, Keogh JM, Henning E, Yeo GS, O'Rahilly S, Froguel P, Farooqi IS, Blakemore AI. A deletion of the HBII-85 class of small nucleolar RNAs (snoRNAs) is associated with hyperphagia, obesity and hypogonadism. *Hum Mol Genet*. 2009 Sep 1;18(17):3257–65
22. Duker AL, Ballif BC, Bawle EV, Person RE, Mahadevan S, Alliman S, Thompson R, Traylor R, Bejjani BA, Shaffer LG, Rosenfeld JA, Lamb AN, Sahoo T. Paternally inherited microdeletion at 15q11.2 confirms a significant role for the SNORD116 C/D box snoRNA cluster in Prader-Willi syndrome. *Eur J Hum Genet*. 2010 Nov;18(11):1196–201
23. Sahoo T, del Gaudio D, German JR, Shinawi M, Peters SU, Person RE, Garnica A, Cheung SW, Beaudet AL. Prader-Willi phenotype caused by paternal deficiency for the HBII-85 C/D box small nucleolar RNA cluster. *Nat Genet*. 2008 Jun;40(6):719–21
24. For a full description see http://omim.org/entry/180860
25. For a full description see http://omim.org/entry/130650
26. Data collated in Kotzot D. Maternal uniparental disomy 14 dissection of the phenotype with respect to rare autosomal recessively inherited traits, trisomy mosaicism, and genomic imprinting. *Ann Genet*. 2004 Jul-Sep;47(3):251–60
27. Kagami M, Sekita Y, Nishimura G, Irie M, Kato F, Okada M, Yamamori S, Kishimoto H, Nakayama M, Tanaka Y, Matsuoka K, Takahashi T, Noguchi M, Tanaka Y, Masumoto K, Utsunomiya T, Kouzan H, Komatsu Y, Ohashi H, Kurosawa K, Kosaki K, Ferguson-Smith AC, Ishino F, Ogata T. Deletions and epimutations affecting the human 14q32.2 imprinted region in individuals with paternal and maternal upd(14)-like phenotypes. *Nat Genet*. 2008 Feb;40(2):237–42
28. For a detailed review of the inheritance and clinical characteristics of various human imprinting disorders, see the review by Ishida M, Moore GE. The role of imprinted genes in humans. *Mol Aspects Med*. 2013 Jul-Aug;34(4):826–40

29. Press release on 14 October 2013 from American Society for Reproductive Medicine http://www.asrm.org/Five_Million_Babies_Born_with_Help_of_Assisted_Reproductive_Technologies/
30. This is discussed in some detail in Ishida M, Moore GE. The role of imprinted genes in humans. *Mol Aspects Med*. 2013 Jul–Aug;34(4):826–40

Chapter 11

1. Reviewed in Moss T, Langlois F, Gagnon-Kugler T, Stefanovsky V. A housekeeper with power of attorney: the rRNA genes in ribosome biogenesis. *Cell Mol Life Sci*. 2007 Jan;64(1):29–49
2. For more information on ribosomes and rRNAs it is easiest to refer to a good molecular biology textbook such as *Molecular Biology of the Cell, 5th Edition* by Alberts, Johnson, Lewis, Raff, Roberts and Walter, 2012.
3. http://www.nobelprize.org/educational/medicine/dna/a/translation/trna.html
4. http://www.bscb.org/?url=softcell/ribo
5. Reviewed in Zentner GE, Saiakhova A, Manaenkov P, Adams MD, Scacheri PC. Integrative genomic analysis of human ribosomal DNA. *Nucleic Acids Res*. 2011 Jul;39(12):4949–60
6. This whole area of diseases caused by defects in ribosomal proteins is interestingly, if occasionally rather provocatively reviewed in Narla A, Ebert BL. Ribosomopathies: human disorders of ribosome dysfunction. *Blood*. 2010 Apr 22;115(16):3196–205
7. International Human Genome Sequencing Consortium. Initial sequencing and analysis of the human genome. *Nature*. 2001 Feb 15;409(6822):860–921
8. See for example Hedges SB, Blair JE, Venturi ML, Shoe JL. A molecular timescale of eukaryote evolution and the rise of complex multicellular life. *BMC Evol Biol*. 2004 Jan 28;4:2
9. Reviewed in Wilson DN. Ribosome-targeting antibiotics and mechanisms of bacterial resistance. *Nat Rev Microbiol*. 2014 Jan;12(1):35–48
10. http://www.genenames.org/rna/TRNA#MTTRNA
11. Once again I would recommend a good molecular biology textbook if you would like to learn more, such as *Molecular Biology of the Cell, 5th Edition* by Alberts, Johnson, Lewis, Raff, Roberts and Walter, 2012

12. McFarland R, Schaefer AM, Gardner JL, Lynn S, Hayes CM, Barron MJ, Walker M, Chinnery PF, Taylor RW, Turnbull DM. Familial myopathy: new insights into the T14709C mitochondrial tRNA mutation. *Ann Neurol.* 2004 Apr;55(4):478–84
13. Zheng J, Ji Y, Guan MX. Mitochondrial tRNA mutations associated with deafness. *Mitochondrion.* 2012 May;12(3):406–13
14. Qiu Q, Li R, Jiang P, Xue L, Lu Y, Song Y, Han J, Lu Z, Zhi S, Mo JQ, Guan MX. Mitochondrial tRNA mutations are associated with maternally inherited hypertension in two Han Chinese pedigrees. *Hum Mutat.* 2012 Aug;33(8):1285–93
15. Giordano C, Perli E, Orlandi M, Pisano A, Tuppen HA, He L, Ierinò R, Petruzziello L, Terzi A, Autore C, Petrozza V, Gallo P, Taylor RW, d'Amati G. Cardiomyopathies due to homoplasmic mitochondrial tRNA mutations: morphologic and molecular features. *Hum Pathol.* 2013 Jul;44(7):1262–70
16. Lincoln TA, Joyce GF. Self-sustained replication of an RNA enzyme. *Science.* 2009 Feb 27;323(5918):1229–32
17. Sczepanski JT, Joyce GF. A cross-chiral RNA polymerase ribozyme. *Nature.* Published online 29 October 2014

Chapter 12

1. An overview of MYC's role, and the importance of chromosomal rearrangements can be found in Ott G, Rosenwald A, Campo E. Understanding MYC-driven aggressive B-cell lymphomas: pathogenesis and classification. *Blood.* 2013 Dec 5;122(24):3884–91
2. http://www.nlm.nih.gov/medlineplus/ency/article/001308.htm
3. Whyte WA, Orlando DA, Hnisz D, Abraham BJ, Lin CY, Kagey MH, Rahl PB, Lee TI, Young RA. Master transcription factors and mediator establish super-enhancers at key cell identity genes. *Cell.* 2013 Apr 11;153(2):307–19
4. Ostuni R, Piccolo V, Barozzi I, Polletti S, Termanini A, Bonifacio S, Curina A, Prosperini E, Ghisletti S, Natoli G. Latent enhancers activated by stimulation in differentiated cells. *Cell.* 2013 Jan 17;152(1–2):157–71
5. Akhtar-Zaidi B, Cowper-Sal-lari R, Corradin O, Saiakhova A, Bartels CF, Balasubramanian D, Myeroff L, Lutterbaugh J, Jarrar A, Kalady

MF, Willis J, Moore JH, Tesar PJ, Laframboise T, Markowitz S, Lupien M, Scacheri PC. Epigenomic enhancer profiling defines a signature of colon cancer. *Science*. 2012 May 11;336(6082):736–9

6. ENCODE Project Consortium, Bernstein BE, Birney E, Dunham I, Green ED, Gunter C, Snyder M. An integrated encyclopedia of DNA elements in the human genome. *Nature*. 2012 Sep 6;489(7414):57–74
7. For a description of these types of long non-coding RNAs see Ørom UA, Shiekhattar R. Long noncoding RNAs usher in a new era in the biology of enhancers. *Cell*. 2013 Sep 12;154(6):1190–3
8. Ørom UA, Derrien T, Beringer M, Gumireddy K, Gardini A, Bussotti G, Lai F, Zytnicki M, Notredame C, Huang Q, Guigo R, Shiekhattar R. Long noncoding RNAs with enhancer-like function in human cells. *Cell*. 2010 Oct 1;143(1):46–58
9. De Santa F, Barozzi I, Mietton F, Ghisletti S, Polletti S, Tusi BK, Muller H, Ragoussis J, Wei CL, Natoli G. A large fraction of extragenic RNA pol II transcription sites overlap enhancers. *PLoS Biol*. 2010 May 11;8(5):e1000384
10. Hah N, Murakami S, Nagari A, Danko CG, Kraus WL. Enhancer transcripts mark active estrogen receptor binding sites. *Genome Res*. 2013 Aug;23(8):1210–23
11. Lai F, Ørom UA, Cesaroni M, Beringer M, Taatjes DJ, Blobel GA, Shiekhattar R. Activating RNAs associate with Mediator to enhance chromatin architecture and transcription. *Nature*. 2013 Feb 28;494(7438):497–501
12. Risheg H, Graham JM Jr, Clark RD, Rogers RC, Opitz JM, Moeschler JB, Peiffer AP, May M, Joseph SM, Jones JR, Stevenson RE, Schwartz CE, Friez MJ. A recurrent mutation in MED12 leading to R961W causes Opitz-Kaveggia syndrome. *Nat Genet*. 2007 Apr;39(4):451–3
13. The role of super-enhancers in pluripotent cells was first identified in Whyte WA, Orlando DA, Hnisz D, Abraham BJ, Lin CY, Kagey MH, Rahl PB, Lee TI, Young RA. Master transcription factors and mediator establish super-enhancers at key cell identity genes. *Cell*. 2013 Apr 11;153(2):307–19
14. Takahashi K, Yamanaka S. Induction of pluripotent stem cells from mouse embryonic and adult fibroblast cultures by defined factors. *Cell*. 2006 Aug 25;126(4):663–76

15. http://www.nobelprize.org/nobel_prizes/medicine/laureates/2012/
16. Lovén J, Hoke HA, Lin CY, Lau A, Orlando DA, Vakoc CR, Bradner JE, Lee TI, Young RA. Selective inhibition of tumor oncogenes by disruption of super-enhancers. *Cell.* 2013 Apr 11;153(2):320–34
17. For an overview of the various molecular causes see Skibbens RV, Colquhoun JM, Green MJ, Molnar CA, Sin DN, Sullivan BJ, Tanzosh EE. Cohesinopathies of a feather flock together. *PLoS Genet.* 2013 Dec;9(12):e1004036
18. http://www.cdls.org.uk/information-centre/
19. Sanyal A, Lajoie BR, Jain G, Dekker J. The long-range interaction landscape of gene promoters. *Nature.* 2012 Sep 6;489(7414):109–13
20. Jackson DA, Hassan AB, Errington RJ, Cook PR. Visualization of focal sites of transcription within human nuclei. *EMBO J.* 1993 Mar;12(3):1059–65
21. For an excellent review of this topic see Rieder D, Trajanoski Z, McNally JG. Transcription factories. *Front Genet.* 2012 Oct 23;3:221. doi: 10.3389/fgene.2012.00221. eCollection 2012
22. Iborra FJ, Pombo A, Jackson DA, Cook PR. Active RNA polymerases are localized within discrete transcription 'factories' in human nuclei. *J Cell Sci.* 1996 Jun;109 (Pt 6):1427–36
23. Jackson DA, Iborra FJ, Manders EM, Cook PR. Numbers and organization of RNA polymerases, nascent transcripts, and transcription units in HeLa nuclei. *Mol Biol Cell.* 1998 Jun;9(6):1523–36
24. Papantonis A, Larkin JD, Wada Y, Ohta Y, Ihara S, Kodama T, Cook PR. Active RNA polymerases: mobile or immobile molecular machines? *PLoS Biol.* 2010 Jul 13;8(7):e1000419
25. Osborne CS, Chakalova L, Brown KE, Carter D, Horton A, Debrand E, Goyenechea B, Mitchell JA, Lopes S, Reik W, Fraser P. Active genes dynamically colocalize to shared sites of ongoing transcription. *Nat Genet.* 2004 Oct;36(10):1065–71
26. Osborne CS, Chakalova L, Mitchell JA, Horton A, Wood AL, Bolland DJ, Corcoran AE, Fraser P. Myc dynamically and preferentially relocates to a transcription factory occupied by Igh. *PLoS Biol.* 2007 Aug;5(8):e192

Chapter 13

1. It's difficult to find a definitive first use of this description, as discussed in http://english.stackexchange.com/questions/103851/where-does-the-phrase-of-boredom-punctuated-by-moments-of-terror-come-from
2. For a review of this, see Moltó E, Fernández A, Montoliu L. Boundaries in vertebrate genomes: different solutions to adequately insulate gene expression domains. *Brief Funct Genomic Proteomic*. 2009 Jul;8(4):283–96
3. Ishihara K, Oshimura M, Nakao M. CTCF-dependent chromatin insulator is linked to epigenetic remodeling. *Mol Cell*. 2006 Sep 1;23(5):733–42
4. Lutz M, Burke LJ, Barreto G, Goeman F, Greb H, Arnold R, Schultheiss H, Brehm A, Kouzarides T, Lobanenkov V, Renkawitz R. Transcriptional repression by the insulator protein CTCF involves histone deacetylases. *Nucleic Acids Res*. 2000 Apr 15;28(8):1707–13
5. Lunyak VV, Prefontaine GG, Núñez E, Cramer T, Ju BG, Ohgi KA, Hutt K, Roy R, García-Díaz A, Zhu X, Yung Y, Montoliu L, Glass CK, Rosenfeld MG. Developmentally regulated activation of a SINE B2 repeat as a domain boundary in organogenesis. *Science*. 2007 Jul 13;317(5835):248–51
6. Reviewed in Kirkland JG, Raab JR, Kamakaka RT. TFIIIC bound DNA elements in nuclear organization and insulation. *Biochim Biophys Acta*. 2013 Mar–Apr;1829(3–4):418–24
7. This is known as Turner's syndrome and more information can be found at http://www.nhs.uk/Conditions/Turners-syndrome/Pages/Introduction.aspx
8. For more information see http://ghr.nlm.nih.gov/condition/triple-x-syndrome
9. This condition is known as Klinefelter's syndrome and more information can be found at http://ghr.nlm.nih.gov/condition/klinefelter-syndrome
10. *Star Trek: First Contact* (1996). By far the best of all the Star Trek movies, at least until the JJ Abrams franchise reboot.
11. See https://ghr.nlm.nih.gov/gene/SHOX
12. Hemani G, Yang J, Vinkhuyzen A, Powell JE, Willemsen G, Hottenga JJ,

Abdellaoui A, Mangino M, Valdes AM, Medland SE, Madden PA, Heath AC, Henders AK, Nyholt DR, de Geus EJ, Magnusson PK, Ingelsson E, Montgomery GW, Spector TD, Boomsma DI, Pedersen NL, Martin NG, Visscher PM. Inference of the genetic architecture underlying BMI and height with the use of 20,240 sibling pairs. *Am J Hum Genet*. 2013 Nov 7;93(5):865–75

Chapter 14

1. A wealth of information about ENCODE, including interviews with some of the leading scientists, can be accessed at http://www.nature.com/encode/
2. http://www.theguardian.com/science/2012/sep/05/genes-genome-junk-dna-encode
3. http://edition.cnn.com/2012/09/05/health/encode-human-genome/index.html?hpt=hp_bn12
4. http://www.telegraph.co.uk/science/science-news/9524165/Worldwide-army-of-scientists-cracks-the-junk-DNA-code.html
5. ENCODE Project Consortium, Bernstein BE, Birney E, Dunham I, Green ED, Gunter C, Snyder M. An integrated encyclopedia of DNA elements in the human genome. *Nature*. 2012 Sep 6;489(7414):57–74
6. Mattick JS. A new paradigm for developmental biology. *J Exp Biol*. 2007 May;210(Pt 9):1526–47
7. Sanyal A, Lajoie BR, Jain G, Dekker J. The long-range interaction landscape of gene promoters. *Nature*. 2012 Sep 6;489(7414):109–13
8. Thurman RE, Rynes E, Humbert R, Vierstra J, Maurano MT, Haugen E, Sheffield NC, Stergachis AB, Wang H, Vernot B, Garg K, John S, Sandstrom R, Bates D, Boatman L, Canfield TK, Diegel M, Dunn D, Ebersol AK, Frum T, Giste E, Johnson AK, Johnson EM, Kutyavin T, Lajoie B, Lee BK, Lee K, London D, Lotakis D, Neph S, Neri F, Nguyen ED, Qu H, Reynolds AP, Roach V, Safi A, Sanchez ME, Sanyal A, Shafer A, Simon JM, Song L, Vong S, Weaver M, Yan Y, Zhang Z, Zhang Z, Lenhard B, Tewari M, Dorschner MO, Hansen RS, Navas PA, Stamatoyannopoulos G, Iyer VR, Lieb JD, Sunyaev SR, Akey JM, Sabo PJ, Kaul R, Furey TS, Dekker J, Crawford GE, Stamatoyannopoulos JA. The accessible chromatin landscape of the human genome. *Nature*. 2012 Sep 6;489(7414):75–82

9. Djebali S, Davis CA, Merkel A, Dobin A, Lassmann T, Mortazavi A, Tanzer A, Lagarde J, Lin W, Schlesinger F, Xue C, Marinov GK, Khatun J, Williams BA, Zaleski C, Rozowsky J, Röder M, Kokocinski F, Abdelhamid RF, Alioto T, Antoshechkin I, Baer MT, Bar NS, Batut P, Bell K, Bell I, Chakrabortty S, Chen X, Chrast J, Curado J, Derrien T, Drenkow J, Dumais E, Dumais J, Duttagupta R, Falconnet E, Fastuca M, Fejes-Toth K, Ferreira P, Foissac S, Fullwood MJ, Gao H, Gonzalez D, Gordon A, Gunawardena H, Howald C, Jha S, Johnson R, Kapranov P, King B, Kingswood C, Luo OJ, Park E, Persaud K, Preall JB, Ribeca P, Risk B, Robyr D, Sammeth M, Schaffer L, See LH, Shahab A, Skancke J, Suzuki AM, Takahashi H, Tilgner H, Trout D, Walters N, Wang H, Wrobel J, Yu Y, Ruan X, Hayashizaki Y, Harrow J, Gerstein M, Hubbard T, Reymond A, Antonarakis SE, Hannon G, Giddings MC, Ruan Y, Wold B, Carninci P, Guigó R, Gingeras TR. Landscape of transcription in human cells. *Nature*. 2012 Sep 6;489(7414):101–8
10. I originally used this description in a Huffington Post blog about the ENCODE project. I've decided I like it so much I will use it again here! For the original blog, see http://www.huffingtonpost.com/nessa-carey/the-value-of-encode_b_1909153.html
11. A good example can be found at http://blog.art21.org/2009/03/06/on-representations-of-the-artist-at-work-part-2/#.UyDZjZZFDIU
12. Ward LD, Kellis M. Evidence of abundant purifying selection in humans for recently acquired regulatory functions. *Science*. 2012 Sep 28;337(6102):1675–8.
13. Ecker JR, Bickmore WA, Barroso I, Pritchard JK, Gilad Y, Segal E. Genomics: ENCODE explained. *Nature*. 2012 Sep 6;489(7414)
14. For a fascinating example of epigenetic transgenerational inheritance see this paper, in which a fear response was passed on from parent to pups: Dias BG, Ressler KJ. Parental olfactory experience influences behavior and neural structure in subsequent generations. *Nat Neurosci*. 2014 Jan;17(1):89–96
15. Graur D, Zheng Y, Price N, Azevedo RB, Zufall RA, Elhaik E. On the immortality of television sets: 'function' in the human genome according to the evolution-free gospel of ENCODE. *Genome Biol Evol*. 2013;5(3):578–90

Chapter 15

1. http://womenshistory.about.com/od/mythsofwomenshistory/a/Did-Anne-Boleyn-Really-Have-Six-Fingers-On-One-Hand.htm
2. Lettice LA, Heaney SJ, Purdie LA, Li L, de Beer P, Oostra BA, Goode D, Elgar G, Hill RE, de Graaff E. A long-range Shh enhancer regulates expression in the developing limb and fin and is associated with preaxial polydactyly. *Hum Mol Genet*. 2003 Jul 15;12(14):1725–35
3. www.hemingwayhome.com/cats/
4. Lettice LA, Hill AE, Devenney PS, Hill RE. Point mutations in a distant sonic hedgehog cis-regulator generate a variable regulatory output responsible for preaxial polydactyly. *Hum Mol Genet*. 2008 Apr 1;17(7):978–85
5. For a fuller description, see http://www.genome.gov/12512735
6. Jeong Y, Leskow FC, El-Jaick K, Roessler E, Muenke M, Yocum A, Dubourg C, Li X, Geng X, Oliver G, Epstein DJ. Regulation of a remote Shh forebrain enhancer by the Six3 homeoprotein. *Nat Genet*. 2008 Nov;40(11):1348–53
7. For more information see http://rarediseases.info.nih.gov/gard/10874/pancreatic-agenesis/resources/1
8. Lango Allen H, Flanagan SE, Shaw-Smith C, De Franco E, Akerman I, Caswell R; International Pancreatic Agenesis Consortium, Ferrer J, Hattersley AT, Ellard S. GATA6 haploinsufficiency causes pancreatic agenesis in humans. *Nat Genet*. 2011 Dec 11;44(1):20–2
9. Sellick GS, Barker KT, Stolte-Dijkstra I, Fleischmann C, Coleman RJ, Garrett C, Gloyn AL, Edghill EL, Hattersley AT, Wellauer PK, Goodwin G, Houlston RS. Mutations in PTF1A cause pancreatic and cerebellar agenesis. *Nat Genet*. 2004 Dec;36(12):1301–5
10. Weedon MN, Cebola I, Patch AM, Flanagan SE, De Franco E, Caswell R, Rodríguez-Seguí SA, Shaw-Smith C, Cho CH, Lango Allen H, Houghton JA, Roth CL, Chen R, Hussain K, Marsh P, Vallier L, Murray A; International Pancreatic Agenesis Consortium, Ellard S, Ferrer J, Hattersley AT. Recessive mutations in a distal PTF1A enhancer cause isolated pancreatic agenesis. *Nat Genet*. 2014 Jan;46(1):61–4
11. For a review of this, see Sturm RA. Molecular genetics of human pigmentation diversity. *Hum Mol Genet*. 2009 Apr 15;18(R1):R9–17

12. Durham-Pierre D, Gardner JM, Nakatsu Y, King RA, Francke U, Ching A, Aquaron R, del Marmol V, Brilliant MH. African origin of an intragenic deletion of the human P gene in tyrosinase positive oculocutaneous albinism. *Nat Genet.* 1994 Jun;7(2):176–9
13. Visser M, Kayser M, Palstra RJ. HERC2 rs12913832 modulates human pigmentation by attenuating chromatin-loop formation between a long-range enhancer and the OCA2 promoter. *Genome Res.* 2012 Mar;22(3):446–55
14. For an up-to-date catalogue, see www.genome.gov/gwastudies/
15. Hindorff LA, Sethupathy P, Junkins HA, Ramos EM, Mehta JP, Collins FS, Manolio TA. Potential etiologic and functional implications of genome-wide association loci for human diseases and traits. *Proc Natl Acad Sci U S A.* 2009 Jun 9;106(23):9362–7
16. Gorkin DU, Ren B. Genetics: Closing the distance on obesity culprits. *Nature.* 2014 Mar 20;507(7492):309–10
17. Frayling TM, Timpson NJ, Weedon MN, Zeggini E, Freathy RM, Lindgren CM, Perry JR, Elliott KS, Lango H, Rayner NW, Shields B, Harries LW, Barrett JC, Ellard S, Groves CJ, Knight B, Patch AM, Ness AR, Ebrahim S, Lawlor DA, Ring SM, Ben-Shlomo Y, Jarvelin MR, Sovio U, Bennett AJ, Melzer D, Ferrucci L, Loos RJ, Barroso I, Wareham NJ, Karpe F, Owen KR, Cardon LR, Walker M, Hitman GA, Palmer CN, Doney AS, Morris AD, Smith GD, Hattersley AT, McCarthy MI. A common variant in the FTO gene is associated with body mass index and predisposes to childhood and adult obesity. *Science.* 2007 May 11;316(5826):889–94
18. Scuteri A, Sanna S, Chen WM, Uda M, Albai G, Strait J, Najjar S, Nagaraja R,Orrú M, Usala G, Dei M, Lai S, Maschio A, Busonero F, Mulas A, Ehret GB, Fink AA,Weder AB, Cooper RS, Galan P, Chakravarti A, Schlessinger D, Cao A, Lakatta E, Abecasis GR. Genome-wide association scan shows genetic variants in the FTO gene are associated with obesity-related traits. *PLoS Genet.* 2007 Jul;3(7):e115
19. Church C, Moir L, McMurray F, Girard C, Banks GT, Teboul L, Wells S, Brüning JC, Nolan PM, Ashcroft FM, Cox RD. Overexpression of Fto leads to increased food intake and results in obesity. *Nat Genet.* 2010 Dec;42(12):1086–92
20. Fischer J, Koch L, Emmerling C, Vierkotten J, Peters T, Brüning JC,

Rüther U. Inactivation of the Fto gene protects from obesity. *Nature*. 2009 Apr 16;458(7240):894–8

21. Smemo S, Tena JJ, Kim KH, Gamazon ER, Sakabe NJ, Gómez-Marín C, Aneas I, Credidio FL, Sobreira DR, Wasserman NF, Lee JH, Puviindran V, Tam D, Shen M, Son JE, Vakili NA, Sung HK, Naranjo S, Acemel RD, Manzanares M, Nagy A, Cox NJ, Hui CC, Gomez-Skarmeta JL, Nóbrega MA. Obesity-associated variants within FTO form long-range functional connections with IRX3. *Nature*. 2014 Mar 20;507(7492):371–5
22. For a recent review of this field see Trent RJ, Cheong PL, Chua EW, Kennedy MA. Progressing the utilisation of pharmacogenetics and pharmacogenomics into clinical care. *Pathology*. 2013 Jun;45(4):357–70
23. http://www.nhs.uk/Conditions/Herceptin/Pages/Introduction.aspx
24. http://www.nature.com/scitable/topicpage/gleevec-the-breakthrough-in-cancer-treatment-565
25. http://www.cancer.gov/cancertopics/druginfo/fda-crizotinib

Chapter 16

1. Examples of such cases can be found at http://medicalmisdiagnosisresearch.wordpress.com/category/osteogenesis-imperfecta-misdiagnosed-as-child-abuse/
2. For a good description of the symptoms and genetics, see http://ghr.nlm.nih.gov/condition/osteogenesis-imperfecta
3. Cho TJ, Lee KE, Lee SK, Song SJ, Kim KJ, Jeon D, Lee G, Kim HN, Lee HR, Eom HH, Lee ZH, Kim OH, Park WY, Park SS, Ikegawa S, Yoo WJ, Choi IH, Kim JW. A single recurrent mutation in the 5′-UTR of IFITM5 causes osteogenesis imperfecta type V. *Am J Hum Genet*. 2012 Aug 10;91(2):343–8
4. Semler O, Garbes L, Keupp K, Swan D, Zimmermann K, Becker J, Iden S, Wirth B, Eysel P, Koerber F, Schoenau E, Bohlander SK, Wollnik B, Netzer C. A mutation in the 5′-UTR of IFITM5 creates an in-frame start codon and causes autosomal-dominant osteogenesis imperfecta type V with hyperplastic callus. *Am J Hum Genet*. 2012 Aug 10;91(2):349–57
5. Moffatt P, Gaumond MH, Salois P, Sellin K, Bessette MC, Godin E, de Oliveira PT, Atkins GJ, Nanci A, Thomas G. Bril: a novel

bone-specific modulator of mineralization. *J Bone Miner Res*. 2008 Sep;23(9):1497–508

6. Liu L, Dilworth D, Gao L, Monzon J, Summers A, Lassam N, Hogg D. Mutation of the CDKN2A 5′ UTR creates an aberrant initiation codon and predisposes to melanoma. *Nat Genet*. 1999 Jan;21(1):128–32
7. Tietze JK, Pfob M, Eggert M, von Preußen A, Mehraein Y, Ruzicka T, Herzinger T. A non-coding mutation in the 5′ untranslated region of patched homologue 1 predisposes to basal cell carcinoma. *Exp Dermatol*. 2013 Dec;22(12):834–5
8. For a full description see http://omim.org/entry/309550
9. Ashley CT Jr, Wilkinson KD, Reines D, Warren ST. FMR1 protein: conserved RNP family domains and selective RNA binding. *Science*. 1993 Oct 22;262(5133):563–6
10. Qin M, Kang J, Burlin TV, Jiang C, Smith CB. Postadolescent changes in regional cerebral protein synthesis: an in vivo study in the FMR1 null mouse. *J Neurosci*. 2005 May 18;25(20):5087–95
11. Azevedo FA, Carvalho LR, Grinberg LT, Farfel JM, Ferretti RE, Leite RE, Jacob Filho W, Lent R, Herculano-Houzel S. Equal numbers of neuronal and nonneuronal cells make the human brain an isometrically scaled-up primate brain. *J Comp Neurol*. 2009 Apr 10;513(5):532–41
12. Drachman DA. Do we have brain to spare? *Neurology*. 2005 Jun 28;64(12):2004–5
13. Darnell JC, Van Driesche SJ, Zhang C, Hung KY, Mele A, Fraser CE, Stone EF, Chen C, Fak JJ, Chi SW, Licatalosi DD, Richter JD, Darnell RB. FMRP stalls ribosomal translocation on messenger RNAs linked to synaptic function and autism. *Cell*. 2011 Jul 22;146(2):247–61
14. Udagawa T, Farny NG, Jakovcevski M, Kaphzan H, Alarcon JM, Anilkumar S, Ivshina M, Hurt JA, Nagaoka K, Nalavadi VC, Lorenz LJ, Bassell GJ, Akbarian S, Chattarji S, Klann E, Richter JD. Genetic and acute CPEB1 depletion ameliorate fragile X pathophysiology. *Nat Med*. 2013 Nov;19(11):1473–7
15. Summarised in http://www.ncbi.nlm.nih.gov/books/NBK1165/
16. Jiang H, Mankodi A, Swanson MS, Moxley RT, Thornton CA. Myotonic dystrophy type 1 is associated with nuclear foci of mutant RNA, sequestration of muscleblind proteins and deregulated alternative splicing in neurons. *Hum Mol Genet*. 2004 Dec 15;13(24):3079–88

17. Savkur RS, Philips AV, Cooper TA. Aberrant regulation of insulin receptor alternative splicing is associated with insulin resistance in myotonic dystrophy. *Nat Genet*. 2001 Sep;29(1):40–7
18. Ho TH, Charlet-B N, Poulos MG, Singh G, Swanson MS, Cooper TA. Muscleblind proteins regulate alternative splicing. *EMBO J*. 2004 Aug 4;23(15):3103–12
19. Kino Y, Washizu C, Oma Y, Onishi H, Nezu Y, Sasagawa N, Nukina N, Ishiura S. MBNL and CELF proteins regulate alternative splicing of the skeletal muscle chloride channel CLCN1. *Nucleic Acids Res*. 2009 Oct;37(19):6477–90
20. Hanson EL, Jakobs PM, Keegan H, Coates K, Bousman S, Dienel NH, Litt M, Hershberger RE. Cardiac troponin T lysine 210 deletion in a family with dilated cardiomyopathy. *J Card Fail*. 2002 Feb;8(1):28–32
21. Reviewed in Michalova E, Vojtesek B, Hrstka R. Impaired pre-messenger RNA processing and altered architecture of 3′ untranslated regions contribute to the development of human disorders. *Int J Mol Sci*. 2013 Jul 26;14(8): 15681–94
22. For a full description of the syndrome see http://ghr.nlm.nih.gov/condition/immune-dysregulation-polyendocrinopathy-enteropathy-x-linked-syndrome
23. Bennett CL, Brunkow ME, Ramsdell F, O'Briant KC, Zhu Q, Fuleihan RL, Shigeoka AO, Ochs HD, Chance PF. A rare polyadenylation signal mutation of the FOXP3 gene (AAUAAA→AAUGAA) leads to the IPEX syndrome. *Immunogenetics*. 2001 Aug;53(6):435–9
24. For further information see http://www.alsa.org/
25. A database of genes believed to be implicated in ALS can be found at http://alsod.iop.kcl.ac.uk/
26. Kwiatkowski TJ Jr, Bosco DA, Leclerc AL, Tamrazian E, Vanderburg CR, Russ C, Davis A, Gilchrist J, Kasarskis EJ, Munsat T, Valdmanis P, Rouleau GA, Hosler BA, Cortelli P, de Jong PJ, Yoshinaga Y, Haines JL, Pericak-Vance MA, Yan J, Ticozzi N, Siddique T, McKenna-Yasek D, Sapp PC, Horvitz HR, Landers JE, Brown RH Jr. Mutations in the FUS/TLS gene on chromosome 16 cause familial amyotrophic lateral sclerosis. *Science*. 2009 Feb 27;323(5918):1205–8
27. Vance C, Rogelj B, Hortobágyi T, De Vos KJ, Nishimura AL, Sreedharan J, Hu X, Smith B, Ruddy D, Wright P, Ganesalingam

J, Williams KL, Tripathi V, Al-Saraj S, Al-Chalabi A, Leigh PN, Blair IP, Nicholson G, de Belleroche J, Gallo JM, Miller CC, Shaw CE. Mutations in FUS, an RNA processing protein, cause familial amyotrophic lateral sclerosis type 6. *Science*. 2009 Feb 27;323(5918):1208–11

28. Lai SL, Abramzon Y, Schymick JC, Stephan DA, Dunckley T, Dillman A, Cookson M, Calvo A, Battistini S, Giannini F, Caponnetto C, Mancardi GL, Spataro R, Monsurro MR, Tedeschi G, Marinou K, Sabatelli M, Conte A, Mandrioli J, Sola P, Salvi F, Bartolomei I, Lombardo F; ITALSGEN Consortium, Mora G, Restagno G, Chiò A, Traynor BJ. FUS mutations in sporadic amyotrophic lateral sclerosis. *Neurobiol Aging*. 2011 Mar;32(3):550.e1–4
29. Sabatelli M, Moncada A, Conte A, Lattante S, Marangi G, Luigetti M, Lucchini M, Mirabella M, Romano A, Del Grande A, Bisogni G, Doronzio PN, Rossini PM, Zollino M. Mutations in the 3′ untranslated region of FUS causing FUS overexpression are associated with amyotrophic lateral sclerosis. *Hum Mol Genet*. 2013 Dec 1;22(23):4748–55

Chapter 17

1. Johnson JM, Castle J, Garrett-Engele P, Kan Z, Loerch PM, Armour CD, Santos R, Schadt EE, Stoughton R, Shoemaker DD. Genome-wide survey of human alternative pre-mRNA splicing with exon junction microarrays. *Science*. 2003 Dec 19;302(5653):2141–4
2. Reviewed in Keren H, Lev-Maor G, Ast G. Alternative splicing and evolution: diversification, exon definition and function. *Nat Rev Genet*. 2010 May;11(5):345–55
3. These steps are laid out very clearly in some reviews e.g. Wang GS, Cooper TA. Splicing in disease: disruption of the splicing code and the decoding machinery. *Nat Rev Genet*. 2007 Oct;8(10):749–61
4. More information on the spliceosome can be found in e.g. Padgett RA. New connections between splicing and human disease. *Trends Genet*. 2012 Apr;28(4):147–54
5. http://ghr.nlm.nih.gov/condition/retinitis-pigmentosa
6. Vithana EN, Abu-Safieh L, Allen MJ, Carey A, Papaioannou M, Chakarova C, Al-Maghtheh M, Ebenezer ND, Willis C, Moore AT,

Bird AC, Hunt DM, Bhattacharya SS. A human homolog of yeast pre-mRNA splicing gene, PRP31, underlies autosomal dominant retinitis pigmentosa on chromosome 19q13.4 (RP11). *Mol Cell.* 2001 Aug;8(2):375–81

7. McKie AB, McHale JC, Keen TJ, Tarttelin EE, Goliath R, van Lith-Verhoeven JJ, Greenberg J, Ramesar RS, Hoyng CB, Cremers FP, Mackey DA, Bhattacharya SS, Bird AC, Markham AF, Inglehearn CF. Mutations in the pre-mRNA splicing factor gene PRPC8 in autosomal dominant retinitis pigmentosa (RP13). *Hum Mol Genet.* 2001 Jul 15;10(15):1555–62
8. Chakarova CF, Hims MM, Bolz H, Abu-Safieh L, Patel RJ, Papaioannou MG, Inglehearn CF, Keen TJ, Willis C, Moore AT, Rosenberg T, Webster AR, Bird AC, Gal A, Hunt D, Vithana EN, Bhattacharya SS. Mutations in HPRP3, a third member of pre-mRNA splicing factor genes, implicated in autosomal dominant retinitis pigmentosa. *Hum Mol Genet.* 2002 Jan 1;11(1):87–92
9. Maita H, Kitaura H, Keen TJ, Inglehearn CF, Ariga H, Iguchi-Ariga SM. PAP-1, the mutated gene underlying the RP9 form of dominant retinitis pigmentosa, is a splicing factor. *Exp Cell Res.* 2004 Nov 1;300(2):283–96
10. Microcephalic osteodysplastic primordial dwarfism type 1 also known as Taybi-Linder syndrome. http://rarediseases.info.nih.gov/gard/5120/microcephalic-osteodysplastic-primordial-dwarfism-type-1/resources/1
11. He H, Liyanarachchi S, Akagi K, Nagy R, Li J, Dietrich RC, Li W, Sebastian N, Wen B, Xin B, Singh J, Yan P, Alder H, Haan E, Wieczorek D, Albrecht B, Puffenberger E, Wang H, Westman JA, Padgett RA, Symer DE, de la Chapelle A. Mutations in U4atac snRNA, a component of the minor spliceosome, in the developmental disorder MOPD I. *Science.* 2011 Apr 8;332(6026):238–40
12. Padgett RA. New connections between splicing and human disease. *Trends Genet.* 2012 Apr;28(4):147–54
13. Haas JT, Winter HS, Lim E, Kirby A, Blumenstiel B, DeFelice M, Gabriel S, Jalas C, Branski D, Grueter CA, Toporovski MS, Walther TC, Daly MJ, Farese RV Jr. DGAT1 mutation is linked to a congenital diarrheal disorder. *J Clin Invest.* 2012 Dec 3;122(12):4680–4
14. Byun M, Abhyankar A, Lelarge V, Plancoulaine S, Palanduz A, Telhan

L, Boisson B, Picard C, Dewell S, Zhao C, Jouanguy E, Feske S, Abel L, Casanova JL. Whole-exome sequencing-based discovery of STIM1 deficiency in a child with fatal classic Kaposi sarcoma. *J Exp Med.* 2010 Oct 25;207(11):2307–12

15. See http://www.genome.gov/11007255
16. Eriksson M, Brown WT, Gordon LB, Glynn MW, Singer J, Scott L, Erdos MR, Robbins CM, Moses TY, Berglund P, Dutra A, Pak E, Durkin S, Csoka AB, Boehnke M, Glover TW, Collins FS. Recurrent de novo point mutations in lamin A cause Hutchinson-Gilford progeria syndrome. *Nature.* 2003 May 15;423(6937):293–8
17. http://www.nhs.uk/conditions/spinal-muscular-atrophy/Pages/Introduction.aspx
18. http://www.smatrust.org/what-is-sma/what-causes-sma/
19. Monani UR, Lorson CL, Parsons DW, Prior TW, Androphy EJ, Burghes AH, McPherson JD. A single nucleotide difference that alters splicing patterns distinguishes the SMA gene SMN1 from the copy gene SMN2. *Hum Mol Genet.* 1999 Jul;8(7):1177–83
20. Cooper TA, Wan L, Dreyfuss G. RNA and disease. *Cell.* 2009 Feb 20;136(4):777–93
21. http://quest.mda.org/news/dmd-drisapersen-outperforms-placebo-walking-test
22. http://www.fiercebiotech.com/story/glaxosmithklines-duchenne-md-drug-mirrors-placebo-effect-phiii/2013-10-07

Chapter 18

1. Ameres SL, Zamore PD. Diversifying microRNA sequence and function. *Nat Rev Mol Cell Biol.* 2013 Aug;14(8):475–88
2. For a more detailed description of classes of smallRNAs, see Castel SE, Martienssen RA. RNA interference in the nucleus: roles for small RNAs in transcription, epigenetics and beyond. *Nat Rev Genet.* 2013 Feb;14(2):100–12
3. Kang SG, Liu WH, Lu P, Jin HY, Lim HW, Shepherd J, Fremgen D, Verdin E, Oldstone MB, Qi H, Teijaro JR, Xiao C. MicroRNAs of the miR-17~92 family are critical regulators of T(FH) differentiation. *Nat Immunol.* 2013 Aug;14(8):849–57
4. Baumjohann D, Kageyama R, Clingan JM, Morar MM, Patel S, de

Kouchkovsky D, Bannard O, Bluestone JA, Matloubian M, Ansel KM, Jeker LT. The microRNA cluster miR-17~92 promotes TFH cell differentiation and represses subset-inappropriate gene expression. *Nat Immunol.* 2013 Aug;14(8):840–8

5. Tassano E, Di Rocco M, Signa S, Gimelli G. De novo 13q31.1-q32.1 interstitial deletion encompassing the miR-17-92 cluster in a patient with Feingold syndrome-2. *Am J Med Genet A.* 2013 Apr;161A(4):894–6
6. For more information see http://ghr.nlm.nih.gov/condition/feingold-syndrome
7. Han YC, Ventura A. Control of T(FH) differentiation by a microRNA cluster. *Nat Immunol.* 2013 Aug;14(8):770–1
8. Reviewed in Koerner MV, Pauler FM, Huang R, Barlow DP. The function of non-coding RNAs in genomic imprinting. *Development.* 2009 Jun;136(11):1771–83
9. Rogler LE, Kosmyna B, Moskowitz D, Bebawee R, Rahimzadeh J, Kutchko K, Laederach A, Notarangelo LD, Giliani S, Bouhassira E, Frenette P, Roy-Chowdhury J, Rogler CE. Small RNAs derived from lncRNA RNase MRP have gene-silencing activity relevant to human cartilage-hair hypoplasia. *Hum Mol Genet.* 2014 Jan 15;23(2):368–82
10. Subramanyam D, Lamouille S, Judson RL, Liu JY, Bucay N, Derynck R, Blelloch R. Multiple targets of miR-302 and miR-372 promote reprogramming of human fibroblasts to induced pluripotent stem cells. *Nat Biotechnol.* 2011 May;29(5):443–8
11. Li Z, Yang CS, Nakashima K, Rana TM. Small RNA-mediated regulation of iPS cell generation. *EMBO J.* 2011 Mar 2;30(5):823–34
12. Ameres SL, Zamore PD. Diversifying microRNA sequence and function. *Nat Rev Mol Cell Biol.* 2013 Aug;14(8):475–88
13. Huang TC, Sahasrabuddhe NA, Kim MS, Getnet D, Yang Y, Peterson JM, Ghosh B, Chaerkady R, Leach SD, Marchionni L, Wong GW, Pandey A. Regulation of lipid metabolism by Dicer revealed through SILAC mice. *J Proteome Res.* 2012 Apr 6;11(4):2193–205
14. Yi R, O'Carroll D, Pasolli HA, Zhang Z, Dietrich FS, Tarakhovsky A, Fuchs E. Morphogenesis in skin is governed by discrete sets of differentially expressed microRNAs. *Nat Genet.* 2006 Mar;38(3):356–62
15. Crist CG, Montarras D, Pallafacchina G, Rocancourt D, Cumano A, Conway SJ, Buckingham M. Muscle stem cell behavior is modified by

microRNA-27 regulation of Pax3 expression. *Proc Natl Acad Sci U S A.* 2009 Aug 11;106(32):13383–7

16. Chen JF, Tao Y, Li J, Deng Z, Yan Z, Xiao X, Wang DZ. microRNA-1 and microRNA-206 regulate skeletal muscle satellite cell proliferation and differentiation by repressing Pax7. *J Cell Biol.* 2010 Sep 6;190(5):867–79
17. da Costa Martins PA, Bourajjaj M, Gladka M, Kortland M, van Oort RJ, Pinto YM, Molkentin JD, De Windt LJ. Conditional dicer gene deletion in the postnatal myocardium provokes spontaneous cardiac remodeling. *Circulation.* 2008 Oct 7;118(15):1567–76
18. de Chevigny A, Coré N, Follert P, Gaudin M, Barbry P, Béclin C, Cremer H. miR-7a regulation of Pax6 controls spatial origin of forebrain dopaminergic neurons. *Nat Neurosci.* 2012 Jun 24;15(8):1120–6
19. Konopka W, Kiryk A, Novak M, Herwerth M, Parkitna JR, Wawrzyniak M, Kowarsch A, Michaluk P, Dzwonek J, Arnsperger T, Wilczynski G, Merkenschlager M, Theis FJ, Köhr G, Kaczmarek L, Schütz G. MicroRNA loss enhances learning and memory in mice. *J Neurosci.* 2010 Nov 3;30(44):14835–42
20. Schaefer A, O'Carroll D, Tan CL, Hillman D, Sugimori M, Llinas R, Greengard P. Cerebellar neurodegeneration in the absence of microRNAs. *J Exp Med.* 2007 Jul 9;204(7):1553–8
21. Pietrzykowski AZ, Friesen RM, Martin GE, Puig SI, Nowak CL, Wynne PM, Siegelmann HT, Treistman SN. Posttranscriptional regulation of BK channel splice variant stability by miR-9 underlies neuroadaptation to alcohol. *Neuron.* 2008 Jul 31;59(2):274–87
22. Hollander JA, Im HI, Amelio AL, Kocerha J, Bali P, Lu Q, Willoughby D, Wahlestedt C, Conkright MD, Kenny PJ. Striatal microRNA controls cocaine intake through CREB signalling. *Nature.* 2010 Jul 8;466(7303):197–202
23. Fernández-Hernando C, Baldán A. MicroRNAs and Cardiovascular Disease. *Curr Genet Med Rep.* 2013 Mar;1(1):30–38
24. For a review, see for example Suzuki H, Maruyama R, Yamamoto E, Kai M. Epigenetic alteration and microRNA dysregulation in cancer. *Front Genet.* 2013 Dec 3;4:258. eCollection 2013
25. Kleinman CL, Gerges N, Papillon-Cavanagh S, Sin-Chan P,

Pramatarova A, Quang DA, Adoue V, Busche S, Caron M, Djambazian H, Bemmo A, Fontebasso AM, Spence T, Schwartzentruber J, Albrecht S, Hauser P, Garami M, Klekner A, Bognar L, Montes L, Staffa A, Montpetit A, Berube P, Zakrzewska M, Zakrzewski K, Liberski PP, Dong Z, Siegel PM, Duchaine T, Perotti C, Fleming A, Faury D, Remke M, Gallo M, Dirks P, Taylor MD, Sladek R, Pastinen T, Chan JA, Huang A, Majewski J, Jabado N. Fusion of TTYH1 with the C19MC microRNA cluster drives expression of a brain-specific DNMT3B isoform in the embryonal brain tumor ETMR. *Nat Genet.* 2014 Jan;46(1):39–44

26. Song SJ, Poliseno L, Song MS, Ala U, Webster K, Ng C, Beringer G, Brikbak NJ, Yuan X, Cantley LC, Richardson AL, Pandolfi PP. MicroRNA-antagonism regulates breast cancer stemness and metastasis via TET-family-dependent chromatin remodeling. *Cell.* 2013 Jul 18;154(2):311–24
27. For an extensive review of this approach, see Schwarzenbach H, Nishida N, Calin GA, Pantel K. Clinical relevance of circulating cell-free microRNAs in cancer. *Nat Rev Clin Oncol.* 2014 Mar;11(3):145–56
28. Chen W, Cai F, Zhang B, Barekati Z, Zhong XY. The level of circulating miRNA-10b and miRNA-373 in detecting lymph node metastasis of breast cancer: potential biomarkers. *Tumour Biol.* 2013 Feb;34(1):455–62
29. Hong F, Li Y, Xu Y, Zhu L. Prognostic significance of serum microRNA-221 expression in human epithelial ovarian cancer. *J Int Med Res.* 2013 Feb;41(1):64–71
30. Shen J, Liu Z, Todd NW, Zhang H, Liao J, Yu L, Guarnera MA, Li R, Cai L, Zhan M, Jiang F. Diagnosis of lung cancer in individuals with solitary pulmonary nodules by plasma microRNA biomarkers. *BMC Cancer.* 2011 Aug 24;11:374
31. For more information see http://emedicine.medscape.com/article/233442-overview
32. Trobaugh DW, Gardner CL, Sun C, Haddow AD, Wang E, Chapnik E, Mildner A, Weaver SC, Ryman KD, Klimstra WB. RNA viruses can hijack vertebrate microRNAs to suppress innate immunity. *Nature.* 2014 Feb 13;506(7487):245–8

33. Jopling CL, Yi M, Lancaster AM, Lemon SM, Sarnow P. Modulation of hepatitis C virus RNA abundance by a liver-specific MicroRNA. *Science*. 2005 Sep 2;309(5740):1577–81

Chapter 19

1. See http://www.fiercepharma.com/special-reports/15-best-selling-drugs-2012 for a summary of the best-selling drugs in recent years
2. There are multiple blogs in this area, for example http://biopharmconsortium.com/rnai-therapeutics-stage-a-comeback
3. More information can be found at http://ghr.nlm.nih.gov/condition/transthyretin-amyloidosis
4. http://investors.alnylam.com/releasedetail.cfm?ReleaseID=805999
5. Updates on this programme can be found at http://mirnarx.com/pipeline/mirna-MRX34.html
6. Koval ED, Shaner C, Zhang P, du Maine X, Fischer K, Tay J, Chau BN, Wu GF, Miller TM. Method for widespread microRNA-155 inhibition prolongs survival in ALS-model mice. *Hum Mol Genet*. 2013 Oct 15;22(20):4127–35
7. Ozsolak F, Kapranov P, Foissac S, Kim SW, Fishilevich E, Monaghan AP, John B, Milos PM. Comprehensive polyadenylation site maps in yeast and human reveal pervasive alternative polyadenylation. *Cell*. 2010 Dec 10;143(6):1018–29
8. A very good review of how antisense expression can regulate genes is Pelechano V, Steinmetz LM. Gene regulation by antisense transcription. *Nat Rev Genet*. 2013 Dec;14(12):880–93
9. http://www.drugs.com/cons/fomivirsen-intraocular.html
10. https://www.bhf.org.uk/heart-matters-online/august-september-2012/medical/familial-hypercholesterolaemia.aspx
11. http://www.medscape.com/viewarticle/804574_5
12. http://www.fda.gov/NewsEvents/Newsroom/PressAnnouncements/ucm337195.htm
13. http://www.medscape.com/viewarticle/781317
14. http://www.nature.com/nrd/journal/v12/n3/full/nrd3963.html
15. Lindow M, Kauppinen S. Discovering the first microRNA-targeted drug. *J Cell Biol*. 2012 Oct 29;199(3):407–12

16. http://www.fiercebiotech.com/story/merck-writes-rnai-punts-sirna-alnylam-175m/2014-01-13
17. http://www.fiercebiotech.com/press-releases/rana-therapeutics-raises-207-million-harness-potential-long-non-coding-rna
18. http://www.bostonglobe.com/business/2014/01/30/dicerna-shares-soar-first-day-trading-after-biotech-raises-million-initial-public-offering/mbwMnXBSPsVCUVkGQLc64I/story.html
19. http://www.dicerna.com/pipeline.php as of 14 April 2014
20. http://www.fiercebiotech.com/story/breaking-novartis-slams-brakes-rnai-development-efforts/2014-04-14

Chapter 20

1. The final story draws together multiple findings from a number of different researchers. Rather than refer to each publication, I recommend the following excellent review article: van der Maarel SM, Miller DG, Tawil R, Filippova GN, Tapscott SJ. Facioscapulohumeral muscular dystrophy: consequences of chromatin relaxation. *Curr Opin Neurol.* 2012 Oct;25(5):614–20
2. This is a distinction, and a terminology, first coined by Sidney Brenner.

Appendix
Human Diseases Cited in the Main Text, in Which Junk DNA Has Been Implicated

Alzheimer's disease May involve over-expression of an antisense RNA that binds to and stabilises the critical BACE1 mRNA.

Angelman syndrome A condition caused by abnormal imprinting. Junk DNA is vital in control of imprinting, including the involvement of imprinting control regions, promoters, long non-coding RNAs and cross-talk with the epigenetic systems.

Aplastic anaemia Around 5 per cent of cases are caused by mutations in some of the critical genes that maintain the lengths of telomeres, the junk regions at the ends of chromosomes.

Basal cell carcinoma A small number of cases are caused by mutations in the non-protein-coding region at the beginning of a gene, which result in decreased expression of the RNA from that gene.

Beckwith-Wiedemann syndrome A condition caused by abnormal imprinting. Junk DNA is vital in control of imprinting, including the involvement of imprinting control regions, promoters, long non-coding RNAs and cross-talk with the epigenetic systems.

Burkitt's lymphoma Caused when the Myc oncogene from chromosome 8 gets translocated to chromosome 14 and placed under the control of the immunoglobulin promoter.

Cancer Junk DNA has been implicated at a number of levels in cancer, such as over-expression of certain long non-coding RNAs in specific cancer types. In most cases, the evidence isn't yet strong enough to determine how significant a role these play in human pathology. However, over-expression of the proteins that maintain the lengths of telomeres, the junk regions at the ends of chromosomes, are now generally accepted as having a causal role in the progression of some tumours. Mis-targeting of epigenetic enzymes to the wrong genes because of abnormal expression of long non-coding RNAs is also under active investigation as another method by which cancer cells proliferate abnormally.

Cartilage-hair hypoplasia Caused by mutations which affect smallRNAs embedded within long non-coding RNAs.

Congenital diarrhoea disorder Caused by a mutation in a splicing signal in a gene.

Cornelia de Lange syndrome Caused by defects in a protein required for the junk-mediated higher-order structuring of DNA.

Down's syndrome Caused by uneven distribution of chromosome 21 to developing gametes, a process dependent on a junk region called the centromere.

Duchenne muscular dystrophy Some cases are caused by mutations which result in abnormal splicing of the dystrophin RNA molecule.

Dyskeratosis congenita Can be caused by mutations in a number of different genes, each of which is involved in maintaining the lengths of telomeres, the junk regions at the ends of chromosomes.

Edward's syndrome Caused by uneven distribution of chromosome 18 to developing gametes, a process dependent on a junk region called the centromere.

ETMR paediatric brain tumour Caused by rearrangement and amplification of a smallRNA cluster.

Extra digits Caused by single base changes in an enhancer for a morphogen.

Facioscapulohumeral muscular dystrophy Caused by the interactions of a combination of junk DNA elements, leading to abnormal expression of a retroviral sequence.

Feingold syndrome Some cases are caused by the loss of a cluster of smallRNAs.

Fragile X syndrome of mental retardation Caused by the expansion of a CCG repeat in a non-protein-coding region at the beginning of a gene. The repeat prevents expression of the gene by making it difficult for the cell to copy the DNA into RNA.

Friedreich's ataxia Caused by the expansion of a GAA repeat in a non-protein-coding region within a gene. The repeat prevents expression of the gene by making it difficult for the cell to copy the DNA into RNA.

Hepatitis C virus A smallRNA produced by human liver cells binds to the viral RNA, stabilising it and promoting viral productivity.

HHV-8 susceptibility Can be caused by mutation in a splicing signal in a gene.

Holoprosencephaly Some cases have been shown to be caused by mutations in an enhancer region for a morphogen.

Hutchinson-Gilford progeria Caused by a mutation which creates an extra splicing signal in a gene.

Idiopathic pulmonary fibrosis Can be caused by mutations in a number of different genes, each of which is involved in maintaining the lengths of telomeres, the junk regions at the ends of chromosomes.

IPEX autoimmune disorder Caused by a mutation in the non-protein-coding region at the end of a gene, which prevents correct processing of the mRNA.

Malignant melanoma A small number of cases are caused by mutations in the non-protein-coding region at the beginning of a gene, which result in the insertion of extra amino acids into the protein.

Myotonic dystrophy Caused by the expansion of a CTG repeat in a non-protein-coding region at the end of a gene. The repeat is copied into RNA, and mops up RNA-binding proteins, resulting in mis-regulation of a large number of other mRNA molecules.

Neuropathic pain May involve over-expression of a long non-coding RNA that regulates expression of a key ion channel.

North American eastern equine encephalitis virus A smallRNA produced by human immune cells binds to the viral genome and prevents the immune system from recognising that the body is under attack.

Ohio Amish dwarfism Caused by a mutation in a non-coding RNA required for the proper functioning of the splicing machinery.

Opitz-Kaveggia syndrome Caused by defects in a protein that is critical for interaction with long non-coding RNAs in the Mediator complex.

Osteogenesis imperfecta (brittle bone disease) A small number of cases are caused by mutations in the non-protein-coding region at the beginning of a gene, which result in the insertion of extra amino acids into the protein.

Pancreatic agenesis Some cases have been shown to be caused by mutations in enhancer sequences.

Patau's syndrome Caused by uneven distribution of chromosome 13 to developing gametes, a process dependent on a junk region called the centromere.

Prader-Willi syndrome A condition caused by abnormal imprinting. Junk DNA is vital in control of imprinting, including the involvement of imprinting control regions, promoters, long non-coding RNAs and cross-talk with the epigenetic systems.

Retinitis pigmentosa Some cases are caused by a defect in a protein which is required to ensure normal splicing and removal of junk DNA from mRNA molecules.

Roberts syndrome Caused by defects in a protein required for the junk-mediated higher-order structuring of DNA.

Silver-Russell syndrome A condition caused by abnormal imprinting. Junk DNA is vital in control of imprinting, including the involvement of imprinting control regions, promoters, long non-coding RNAs and cross-talk with the epigenetic systems.

Spinal muscular atrophy The SMN2 gene is unable to compensate for mutations in the closely related SMN1 gene, because of a variant base pair which prevents normal splicing of SMN2 mRNA into functional protein.

X0 syndrome (Turner's syndrome) Women with only one X chromosome, caused by uneven distribution of X chromosomes to developing gametes, a process dependent on a junk region called the centromere.

XXX syndrome Women with three X chromosomes, caused by uneven distribution of X chromosomes to developing gametes, a process dependent on a junk region called the centromere.

XXY syndrome (Klinefelter's syndrome) Men with two X chromosomes, caused by uneven distribution of X chromosomes to developing gametes, a process dependent on a junk region called the centromere.

Index